# Smart Composites and Processing

# Smart Composites and Processing

Editor

**Kwang-Jea Kim**

MDPI • Basel • Beijing • Wuhan • Barcelona • Belgrade • Manchester • Tokyo • Cluj • Tianjin

*Editor*
Kwang-Jea Kim
DTR VMS Italy S.r.l.
Brescia
Italy

*Editorial Office*
MDPI
St. Alban-Anlage 66
4052 Basel, Switzerland

This is a reprint of articles from the Special Issue published online in the open access journal *Polymers* (ISSN 2073-4360) (available at: www.mdpi.com/journal/polymers/special_issues/Smart_Compos_Process).

For citation purposes, cite each article independently as indicated on the article page online and as indicated below:

LastName, A.A.; LastName, B.B.; LastName, C.C. Article Title. *Journal Name* **Year**, *Volume Number*, Page Range.

ISBN 978-3-0365-6052-6 (Hbk)
ISBN 978-3-0365-6051-9 (PDF)

© 2022 by the authors. Articles in this book are Open Access and distributed under the Creative Commons Attribution (CC BY) license, which allows users to download, copy and build upon published articles, as long as the author and publisher are properly credited, which ensures maximum dissemination and a wider impact of our publications.

The book as a whole is distributed by MDPI under the terms and conditions of the Creative Commons license CC BY-NC-ND.

# Contents

About the Editor . . . . . . . . . . . . . . . . . . . . . . . . . . . . . . . . . . . . . . . . . . . . . . . . . vii

**Kwang-Jea Kim**
Smart Composites and Processing
Reprinted from: *Polymers* 2022, 14, 4166, doi:10.3390/polym14194166 . . . . . . . . . . . . . . . . 1

**Sangram P. Bhoite, Jonghyuck Kim, Wan Jo, Pravin H. Bhoite, Sawanta S. Mali and Kyu-Hwan Park et al.**
Understanding the Influence of Gypsum upon a Hybrid Flame Retardant Coating on Expanded Polystyrene Beads
Reprinted from: *Polymers* 2022, 14, 3570, doi:10.3390/polym14173570 . . . . . . . . . . . . . . . . 3

**Chang Seok Ryu and Kwang-Jea Kim**
Interfacial Adhesion in Silica-Silane Filled NR Composites: A Short Review
Reprinted from: *Polymers* 2022, 14, 2705, doi:10.3390/polym14132705 . . . . . . . . . . . . . . . . 15

**Jincheol Kim, Jaewon Lee, Sosan Hwang, Kyungjun Park, Sanghyun Hong and Seojin Lee et al.**
Simultaneous Effects of Carboxyl Group-Containing Hyperbranched Polymers on Glass Fiber-Reinforced Polyamide 6/Hollow Glass Microsphere Syntactic Foams
Reprinted from: *Polymers* 2022, 14, 1915, doi:10.3390/polym14091915 . . . . . . . . . . . . . . . . 29

**Kyung-Soo Sung, So-Yeon Kim, Min-Keun Oh and Namil Kim**
Thermal and Adhesion Properties of Fluorosilicone Adhesives Following Incorporation of Magnesium Oxide and Boron Nitride of Different Sizes and Shapes
Reprinted from: *Polymers* 2022, 14, 258, doi:10.3390/polym14020258 . . . . . . . . . . . . . . . . 43

**Min-Su Heo, Tae-Hoon Kim, Young-Wook Chang and Keon Soo Jang**
Near-Infrared Light-Responsive Shape Memory Polymer Fabricated from Reactive Melt Blending of Semicrystalline Maleated Polyolefin Elastomer and Polyaniline
Reprinted from: *Polymers* 2021, 13, 3984, doi:10.3390/polym13223984 . . . . . . . . . . . . . . . . 57

**Tao Zhang and Ho-Jong Kang**
Enhancement of the Processability and Properties of Nylon 6 by Blending with Polyketone
Reprinted from: *Polymers* 2021, 13, 3403, doi:10.3390/polym13193403 . . . . . . . . . . . . . . . . 67

**Sung-Hun Lee, Su-Yeol Park, Kyung-Ho Chung and Keon-Soo Jang**
Phlogopite-Reinforced Natural Rubber (NR)/Ethylene-Propylene-Diene Monomer Rubber (EPDM) Composites with Aminosilane Compatibilizer
Reprinted from: *Polymers* 2021, 13, 2318, doi:10.3390/polym13142318 . . . . . . . . . . . . . . . . 83

**Hye-Seon Park and Chang-Kook Hong**
Anion Exchange Membrane Based on Sulfonated Poly (Styrene-Ethylene-Butylene-Styrene) Copolymers
Reprinted from: *Polymers* 2021, 13, 1669, doi:10.3390/polym13101669 . . . . . . . . . . . . . . . . 97

**Seo-Hwa Hong, Jin Hwan Park, Oh Young Kim and Seok-Ho Hwang**
Preparation of Chemically Modified Lignin-Reinforced PLA Biocomposites and Their 3D Printing Performance
Reprinted from: *Polymers* 2021, 13, 667, doi:10.3390/polym13040667 . . . . . . . . . . . . . . . . 111

**Tao Zhang, Seung-Jun Lee, Yong Hwan Yoo, Kyu-Hwan Park and Ho-Jong Kang**
Compression Molding of Thermoplastic Polyurethane Foam Sheets with Beads Expanded by Supercritical $CO_2$ Foaming
Reprinted from: *Polymers* **2021**, *13*, 656, doi:10.3390/polym13040656 . . . . . . . . . . . . . . . . . **121**

# About the Editor

**Kwang-Jea Kim**

K. J. Kim received his B.S. and M.S. in Chemical Engineering at INHA U. in S. Korea and Ph.D. in Polymer Engineering at the University of Akron, USA. He was then a Post-Doctoral Fellow at Institute of Polymer Engineering (IPE) at U. Akron. Then, he joined to Struktol Company of America as a Research Scientist/Project Manager of Polymer Processing and Additives R&D. Next, he joined West Virginia University in Chemical Engineering as a Research Assistant Professor. He then worked at the Gyeongsang Nat'l U. in Polymer Sci. and Eng. Dept. as a Brain Pool Research Professor. He then joined at DTR Co. He then joined DTR VMS Italy S.r.l. in the Material R&D Division as Head/Vice President (2019–2020). His professional interests are thermoplastic and rubber compounds, polymer nanocomposites, interfacial science, reactive processing, rheology, morphology, structure–property relationships, chemical additives, nano materials, and synthesis (organic and inorganic hybrid materials). He published three Books (English (2) and Chinese (1)) (1) "Thermoplastic and Rubber Compounds: Technology and Physical Chemistry", Hanser Publisher, Munich, Cincinnati [Co-author with J. L. White] (2008); (2) "Polymer Nanocomposites Handbook" CRC Press, Boca Raton, [Co-author with Rakesh K. Gupta and Elliot Kennel] (2009); (3) [Annotated version of "Polymer Nanocomposites Handbook" CRC Press].

*Editorial*

# Smart Composites and Processing

Kwang-Jea Kim

DTR VMS Italy S.r.l., Via S. Antonio n. 59, Passirano, 25050 Brescia, Italy ; kwangjea.kim@gmail.com

Polymer composites have been at the forefront of research in recent decades as a result of the unique properties they provide for utilization in numerous applications.

The factors affecting the manufacturing of a smart composite includes the choice of ingredients such as polymer, filler and additives as well as their unique composition [1]. These include polymers (modification, blending, general/engineering/super engineering plastic), rubber (natural rubber (NR), EPDM, silicone, TPE, specialty rubber), fibers (size, shape, dispersion/distribution, reinforcement, electrical/thermal conduction, biodegradable, cellulose-, glass-, carbon-, graphene-, aramid-, etc.), and additives (synergistic/antagonistic, anti-degradation, interfacial-adhesion, silane hybrid composites) [2,3]. An example of interfacial adhesion is bifunctional silane [e.g., bis(triethoxysilyilpropyl)disulfide (TESPT) and bis(triethoxysilyilpropyl)tetrasulfide (TESPD)] in a silica-filled rubber system. In the silica filled system, bifunctional silane chemically bonds silica and NR, and not only assists in the dispersion of silica agglomerates in the rubber chain [4,5], but also increases the interfacial interaction between the two materials by chemical bonding [6].

Figure 1 shows the TESPT effect on chemical bonding between silica and NR. Silanes represent smart materials, which change the properties of rubber and plastic composites significantly, as shown above. It was applied for smart/green tire composites, anti-vibration rubber composites, and plastic composites [7,8].

 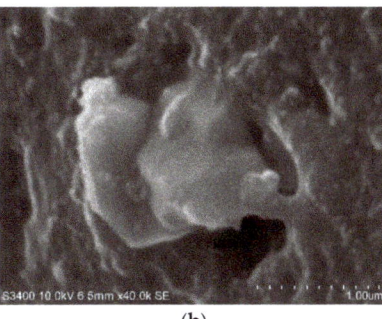

(**a**) (**b**)

**Figure 1.** SEM photograph of silica-NR compound (**a**) without silane, (**b**) with silane (TESPT) [6].

Additionally, the smart processing of polymer composites improves the construction of polymer composites, which is influenced by the choice of mixers, processing condition, processing technique, use of a 3D printer, etc. These include the mixing mechanisms [mixer (internal/open), intermeshing/tangential type (rotor type and screw configuration), reactive mixing (temperature/speed at each stage, curing condition (pre/post), mixing sequence, mold design and molding technique (filler orientation, simulation, pre-treatment (chemical, thermal . . . )) etc.] [4,5,9–13].

These contribute to the construction of advanced polymer composites for high-performance automotive and aerospace parts, advanced electronic devices, environmentally friendly goods, sensors, and other such uses.

"Smart Composites and Processing" is a newly opened Special Issue (SI) of Polymers, which aims to publish original and review papers on the new scientific and applied research and make boundless contributions to the findings and understanding of various smart composites and smart processing.

**Funding:** This research received no external funding.

**Conflicts of Interest:** The authors declare no conflict of interest.

## References

1. White, J.L.; Kim, K.J. *Thermoplastic and Rubber Compounds Technology and Physical Chemistry*; Hanser Publisher: Munich, Germany; Cincinnati, OH, USA, 2008.
2. Plueddemann, E.P. *Silane Coupling Agents*; Plenum Press: New York, NY, USA, 1982.
3. Gupta, R.K.; Kennal, E.; Kim, K.J. *Polymer Nanocomposites Handbook*; CRC Press: Boca Raton, FL, USA, 2009.
4. Kim, K.J.; White, J.L. Silica Agglomerate Breakdown in Three-Stage Mix Including a Continuous Ultrasonic Extruder. *J. Ind. Eng. Chem.* **2000**, *6*, 372–379.
5. Kim, K.J.; White, J.L. TESPT and Different Aliphatic Silane Treated Silica Compounds Effects on Silica Agglomerate Dispersion and on Processability During Mixing in EPDM. *J. Ind. Eng. Chem.* **2001**, *7*, 50–57.
6. Kim, K.J. Silane effects on in-rubber silica dispersion and silica structure ($\alpha$(F)): A Review. *Asian J. Chem.* **2013**, *25*, 5119–5123. [CrossRef]
7. Ryu, C.S.; Kim, K.J. Interfacial Adhesion in Silica-Silane Filled NR Composites: A Short Review. *Polymers* **2022**, *14*, 2705–2719. [CrossRef] [PubMed]
8. Lee, J.Y.; Kim, K.J. MEG effects on hydrolysis of polyamide 66/glass fiber composites and mechanical property changes. *Molecules* **2019**, *24*, 755–765. [CrossRef] [PubMed]
9. Wolff, S. Reinforcing and Vulcanization Effects of Silane Si69 in Silica-Filled Compounds. *Kautsch. Gummi Kunstst.* **1981**, *34*, 280–284.
10. Wolff, S. Optimization of Silane-Silica OTR Compounds. Part 1: Variations of Mixing Temperature and Time during the Modification of Silica with Bis-(3-Triethoxisilylpropyl)-Tetrasulfide. *Rubber Chem. Technol.* **1982**, *55*, 967–989. [CrossRef]
11. Isayev, A.I.; Hong, C.K.; Kim, K.J. Continuous Mixing and Compounding of Polymer/Filler and Polymer/Polymer Mixtures with the Aid of Ultrasound. *Rubber Chem. Technol.* **2003**, *76*, 923–947. [CrossRef]
12. Kim, K.J.; VanderKooi, J. Reactive Batch Mixing for Improved Silica-Silane Coupling. *Int. Polym. Process.* **2004**, *19*, 364–373. [CrossRef]
13. Kim, S.M.; Cho, H.W.; Kim, J.W.; Kim, K.J. Effects of processing geometry on the mechanical properties and silica dispersion of silica-filled isobutylene-isoprene rubber (IIR) compounds. *Elastomers Compos.* **2010**, *45*, 223–229.

## Short Biography of Author

**K. J. Kim** received his M.S. in Chemical Engineering in 1987 at INHA U. in S. Korea and Ph.D. in Polymer Engineering in 1998 at the University of Akron, USA. He was then a Post-Doctoral Fellow at Institute of Polymer Engineering (IPE) at U. Akron (1997–1999). Then, he joined to Struktol Company of America as a Research Scientist/Project Manager of Polymer Processing and Additives R&D (2000–2005). Next, he joined West Virginia University in Chemical Engineering as a Research Assistant Professor (2006–2007). He then worked at the Gyeongsang Nat'l U. in Polymer Sci. and Eng. Dept. as a Brain Pool Research Professor, after being invited by the Korean Government (2008–2009). He then joined Dong Ah Tire & Rubber Co., Ltd. (one division of DTR Co.) as a Chief Technical Officer and moved to Polymer R&D Center at DTR Co. as an R&D Head (2010–2019). He then joined DTR VMS Italy S.r.l. in the Material R&D Division as Head/Vice President (2019–2020). His professional interests are thermoplastic and rubber compounds, polymer nanocomposites, interfacial science, reactive processing, rheology, morphology, structure-property relationships, chemical additives, nano materials, and synthesis (organic and inorganic hybrid materials). He published three Books (English (2) and Chinese (1)) (1) "Thermoplastic and Rubber Compounds: Technology and Physical Chemistry", Hanser Publisher, Munich, Cincinnati [Co-author with J. L. White] (2008), (2) "Polymer Nanocomposites Handbook" CRC Press, Boca Raton, [Co-author with Rakesh K. Gupta and Elliot Kennel] (2009), (3) [Annotated version of "Polymer Nanocomposites Handbook" CRC Press], 78 peer review papers, and filed 11 patents.

Article

# Understanding the Influence of Gypsum upon a Hybrid Flame Retardant Coating on Expanded Polystyrene Beads

Sangram P. Bhoite [1,†], Jonghyuck Kim [2,†], Wan Jo [2], Pravin H. Bhoite [3], Sawanta S. Mali [1], Kyu-Hwan Park [2,*] and Chang Kook Hong [1,*]

[1] School of Chemical Engineering, Chonnam National University, Gwangju 61186, Korea
[2] HDC HYUNDAI EP R & D Center, Yongin-si 16889, Korea
[3] Department of Chemistry, Kisan Veer Mahavidyalaya, Wai 412803, Maharashtra, India
\* Correspondence: kyu@hdc-hyundaiep.com (K.-H.P.); hongck@chonnam.ac.kr (C.K.H.)
† These authors contributed equally to this work.

**Abstract:** A low-cost and effective flame retarding expanded polystyrene (EPS) foam was prepared herein by using a hybrid flame retardant (HFR) system, and the influence of gypsum was studied. The surface morphology and flame retardant properties of the synthesized flame retardant EPS were characterized using scanning electron microscopy (SEM) and cone calorimetry testing (CCT). The SEM micrographs revealed the uniform coating of the gypsum-based HFR on the EPS microspheres. The CCT and thermal conductivity study demonstrated that the incorporation of gypsum greatly decreases the peak heat release rate (PHRR) and total heat release (THR) of the flame retarding EPS samples with acceptable thermal insulation performance. The EPS/HFR with a uniform coating and the optimum amount of gypsum provides excellent flame retardant performance, with a THR of 8 MJ/m$^2$, a PHRR of 53.1 kW/m$^2$, and a fire growth rate (FIGRA) of 1682.95 W/m$^2$s. However, an excessive amount of gypsum weakens the flame retardant performance. The CCT results demonstrate that a moderate gypsum content in the EPS/HFR sample provides appropriate flame retarding properties to meet the fire safety standards.

**Keywords:** hybrid flame retardant materials; influence of gypsum; minimum total heat release

## 1. Introduction

Fire safety via the use of insulating materials is of prime priority in secure building construction. In the last two decades, expanded polystyrene foam (EPS) has become one of the main products in the insulation market due to its moisture resistance, good chemical resistance, and excellent thermal insulation [1–4]. Nevertheless, the highly flammable nature of EPS foam limits its application in the construction industry [5,6]. In recent years, many serious fire tragedies have resulted from the poor flame retardation properties of EPS foam. This represents a serious threat to civilian lives [7,8]. Therefore, it is an immense challenge for industries and researchers to boost the flame resisting properties of EPS foam. Nowadays, researchers focus on the incorporation of various flame retarding materials onto the EPS foam in order to enhance its fire retarding performance, with halogen-free flame retardants now being widely used in the academic and industrial sectors [9].

Among the various flame-retardant additives, intumescent flame retardants (IFRs) are widely used due to their environmental friendliness, low smoke production, and nontoxic properties [10,11]. It is well known that multiple phenomena occur during the combustion of polymeric materials. Thus, during the combustion of EPS foam, the IFR can generate a homogeneous protective char layer that both acts as a barrier to oxygen and heat, and suppresses smoke production, thereby enhancing the flame-retardant capability of the underlying materials [12–15]. In previous work, we prepared a flame-retardant expanded polystyrene foam, and found that, during combustion, the IFR material produced an expanded char layer which acted as an insulating barrier to inhibit heat transfer [16].

Therefore, the formation of an effective and continuous protective char layer is regarded as important for boosting flame retardancy. However, traditional IFR additives are less efficient than halogen-based flame retardants and require significant loadings in order to meet the desirable flammability standards [17]. To overcome this issue, studies suggest that the combination of multiple flame-retardant elements to achieve a synergistic effect would be the best choice [18]. Nevertheless, halogenated flame retardants are still the most efficient flame-retardant materials, and while they may cause environmental issues in some situations, there remains no promising alternative. For example, the bromine-containing molecule decabromodiphenyl ethane (DBDPE) greatly facilitates the gas phase activity of this flame retardant. Several studies suggested that DBDPE does not release the toxic and carcinogenic polybrominated dibenzo-p-dioxin (PBDD) and polybrominated dibenzofuran (PBDF) gases during combustion due to the absence of ether linkages [19–24].

Meanwhile, numerous studies demonstrated that an outstanding flame-retardant performance can be achieved by combining the flame-retardant additives with inorganic flame-retardant fillers, thereby decreasing the proportion of combustible polymers present [25–27]. Moreover, while the addition of a single filler is often less efficient, and does not meet the requisite flammability standards, studies suggest that the combination of multiple mineral fillers can greatly facilitate the flame retarding performance [28,29]. Presently, talc and calcium carbonate ($CaCO_3$) are widely established as flame retardant fillers due to their affordability and thermal stability [30,31]. However, gypsum has attracted particular attention due to its environmental friendliness, cost-effectiveness, thermal stability, and excellent fire resistance [32–36]. Pure gypsum, also known as calcium sulfate dihydrate ($CaSO_4 \cdot 2H_2O$), occurs naturally in crystal form with two water molecules in the crystalline structure. When the gypsum is exposed to heat, these water molecules are gradually released, thereby decreasing the temperature of the polymer matrix and reducing the oxygen concentration. Hence, the dehydrated calcium sulphate formed during combustion of the composite material settles onto the surface to form a protective layer of noncombustible material, thereby greatly contributing to the formation of a fire-resistant barrier against the transfer of heat and gas [37]. Moreover, a polymer binder can be incorporated in the composite material in order to consolidate the flame-retardant ingredients. In this respect, the industrial process is presently focused on the development of water-based formulations due to environmental concerns [38]. Hence, ethylene vinyl acetate emulsion (EVA) is presently used as a binder in the preparation of water-based formulations due to its good adhesion capacity and low-cost.

The present work examined the influence of gypsum upon a novel hybrid flame retardant (HFR) that incorporates ammonium polyphosphate (APP), pentaerythritol (PER), decabromodiphenyl ethane (DBDPE), expandable graphite (EG), calcium carbonate ($CaCO_3$), and talc, which is applied onto the expanded polystyrene (EPS) foam. The thermal performance and flame retardancy of the as-fabricated EPS foam were investigated via a thermogravimetric analysis (TGA) and the cone calorimetry test (CCT). The results indicate that the optimized gypsum-based HFR plays a key role in boosting the flame resistance properties of the EPS foam, with a total heat release (THR) of 8 $MJ/m^2$, a peak heat release rate (PHRR) of 53.1 $kW/m^2$, and a fire growth rate (FIGRA) of 1682.95 $W/m^{-2}s$. In addition, scanning electron microscopy (SEM), and energy dispersive spectroscopy (EDS) were used to investigate the combustion behavior of the residual char. To the best of the authors' knowledge, this is the first time that the coating of EPS foam with a gypsum-based HFR material has been reported for improved flame-retardant performance.

## 2. Materials and Methods
### 2.1. Materials

The expanded polystyrene (EPS) beads, ammonium polyphosphate (APP, purity > 98%), pentaerythritol (PER, purity 98%), calcium carbonate ($CaCO_3$, purity > 98.5%), decabromodiphenyl ethane (DBDPE, purity 99%), talc (whiteness: 94.0 ± 1%, particle size: 11.0 ± 2 μm), and EVA emulsion (G3, solid content 56.5%,) were obtained from HDC

Hyundai EP Co., (Seoul, Korea). The expandable graphite (EG, purity: 99%, size 270 μm) was purchased from Yuil Chemi Tech Co. Ltd., (Seoul, Korea). The gypsum (purity > 96%) was provided by Namhae Chemical Corporation, (Yeosu, Korea).

*2.2. Preparation of the Gypsum-Based HFR Formulation*

The gypsum-based HFR materials were prepared according to the parameters listed in Table 1. In brief, a fixed amount of binder 55 g; (APP:PER:DBDPE:$CaCO_3$ = 15:5:5:5 by mass) was added to 40 g of EG and 0, 9, 12, or 15 g of gypsum in 95, 104, 110, or 114 mL of distilled water, and stirred at room temperature for 48 h to obtain the flame-retardant solutions labelled as HFR0, HFR9, HFR12, and HFR15, respectively.

Table 1. The preparative parameters of the gypsum-based HFR formulations.

| Sample | Binder [a] (g) | Gypsum (g) | EG (g) | Water (mL) |
| --- | --- | --- | --- | --- |
| HFR0 | 55 | 0 | 40 | 95 |
| HFR9 | 55 | 9 | 40 | 104 |
| HFR12 | 55 | 12 | 40 | 110 |
| HFR15 | 55 | 15 | 40 | 114 |

[a] Hybrid flame retardant additive composition was used as APP:PER:DBDPE:$CaCO_3$ =15:5:5:5 by mass.

*2.3. Preparation of the Flame-Retardant EPS Foam*

A simple mixing method was used for the preparation of the flame-retardant EPS foam. When performing the typical procedure, EPS microspheres (13 g) and a HFR sample were mixed in a 1:3 ratio, and the uniformly coated EPS spheres were then transferred into a cuboid mold and hot pressed at 90 °C for 6 h. The cured cuboid EPS foam with the dimensions of $100 \times 100 \times 50$ mm$^3$ was carefully removed from the mold and dried in an oven at 60 °C for 24 h. The EPS samples that were prepared using HFR0, HFR9, HFR12, and HFR15 were correspondingly labelled as EPS, EPS/HFR0 EPS/HFR9, EPS/HFR12, and EPS/HFR15.

*2.4. Characterization*

The surface morphologies of the various EPS/HFR samples and the char residues obtained after combustion were recorded using a scanning electron microscope (SEM; S-4700, Hitachi). Thermogravimetric analyses (TGA; TA Instruments SDTA 851E) of the EPS and EPS/HFR samples were performed at a heating rate of 10 °C/min from room temperature (RT) to 800 °C under a nitrogen atmosphere. The flame-retardant behavior was measured by applying a butane spray gun jet at a distance of 5 cm from the cuboid HFR/EPS sample for complete combustion and the combustion test was monitored by recording digital photographs. To assess the flammability behaviors of the samples in a real fire, their peak heat release rate (PHRR), total heat release (THR), and fire growth rate (FIGRA) were evaluated via the cone calorimetry test (CCT) using a standard cone calorimeter (Fire Testing Technology Limited, UK) according to the ISO5660 standard under an external heat flux of 50 kW/m$^2$ for 600 s. The thermal conductivity coefficient of the neat EPS and flame-retardant EPS foam was measured with a thermal conductivity analyzer (Dow chemical, Yeosu-si, Korea Ltd.). The specimen dimensions were $200 \times 200 \times 20$ mm$^3$.

## 3. Results and Discussion

*3.1. Thermogravimetric Analysis (TGA)*

The thermal degradation behaviors of the various EPS samples are illustrated by the TGA curves in Figure 1. Here, the neat EPS clearly exhibited one-stage decomposition, with 100% weight loss taking place in the temperature range of 350 to 450 °C, so that no char residue remained after thermal decomposition [39]. By comparison, the hybrid EPS/HFR0 sample exhibited a significantly lower initial decomposition temperature, with a major weight loss between 180 and 460 °C due to the early decomposition of the EG and the increasing reaction between the flame-retardant additives, which led to a certain amount of residue char remaining at 800 °C. In detail, when the temperature rises above 180 °C, the EG begins to decompose and release sulfur dioxide [40,41], while the APP component of the binder begins to decompose to release water and ammonia; these decomposition products react to form polyphosphoric acid. Further, at temperatures between 200 and 300 °C, the polyphosphoric acid reacts with the hydroxyl group of the PER component of the binder to form a phosphate ester, which leads to the formation of char [42]. At 300–440 °C, however, the DBDPE component of the binder begins to decompose and release bromine radicals, which react to decrease the oxygen concentration and accelerate the gas phase, thereby boosting the flame retardancy [43–45]. Above 440 °C, an additional weight loss was observed due to the reaction of the talc and calcium carbonate components of the binder with the polyphosphate network [46–48]. With the addition of gypsum ($CaSO_4 \cdot 2H_2O$), the initial decomposition temperature decreased slightly relative to that of EPS/HFR0 due to dehydration of the gypsum to form thermally stable calcium sulfate ($CaSO_4$) [49]. This led to an increase in the final residual weight from 0% for the neat EPS and 19.30% for the EPS/HFR0 to 27.46, 27.80, and 30.52% for the EPS/HFR9, EPS/HFR15, and EPS/HFR12, respectively. These results suggest that the thermally stable $CaSO_4$ interacts with HFR to form a thermally stable char layer structure, which acts as an effective barrier against heat and mass transfer during the combustion process, thereby enhancing the flame-retardant performance of the EPS/HFR foam.

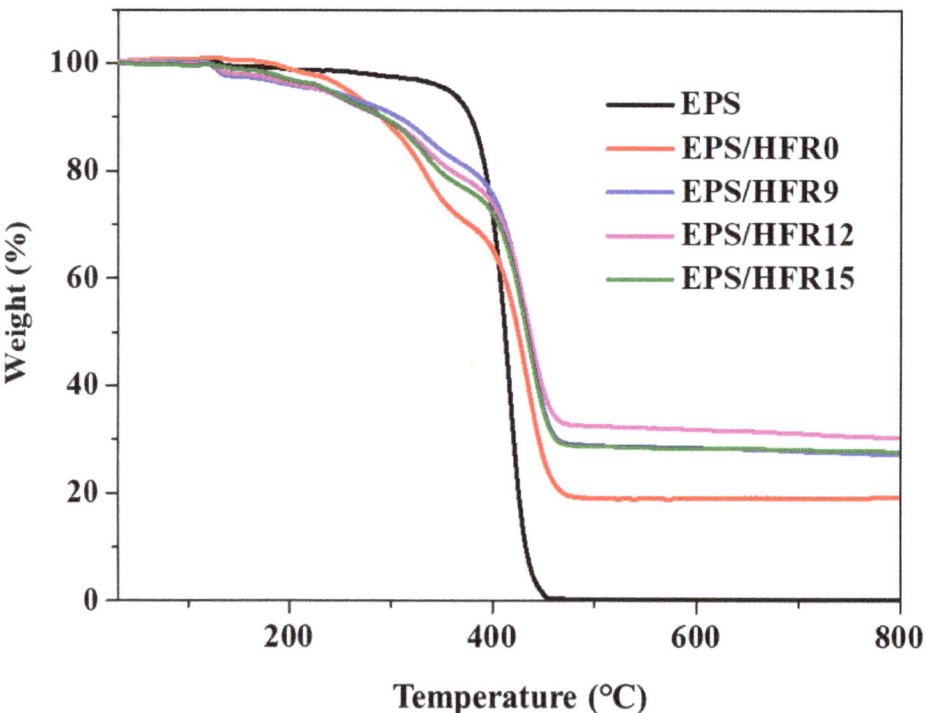

**Figure 1.** The TGA curves of the neat EPS and flame-retardant EPS samples obtained under a nitrogen atmosphere at a heating rate of 10 °C/min.

*3.2. Microstructural Study*

The surface morphologies of the EPS before and after the application of the flame-retardant coating are revealed by the SEM images in Figure 2. Here, the neat EPS exhibits a spherical shape with a very smooth surface morphology (Figure 2a). Further, the cross-sectional image in Figure 2b clearly shows the absence of any coating between the neat EPS beads. Due to its chemical composition, the EPS undergoes a radical chain reaction during combustion, thereby generating volatile products that can act as fuels for the production of toxic black smoke. By contrast, the SEM image of the EPS/HFR12 sample in Figure 2c confirms the successful coating of the EPS microsphere with the gypsum-based HFR materials, and the cross-sectional image in Figure 2d reveals the formation of the gypsum-based HFR coating between two adjacent EPS beads. During combustion, these flame-retardant coating materials can generate a compact char layer that can act as an effective fireproofing barrier, thereby improving the flame resistance performance of the EPS foam.

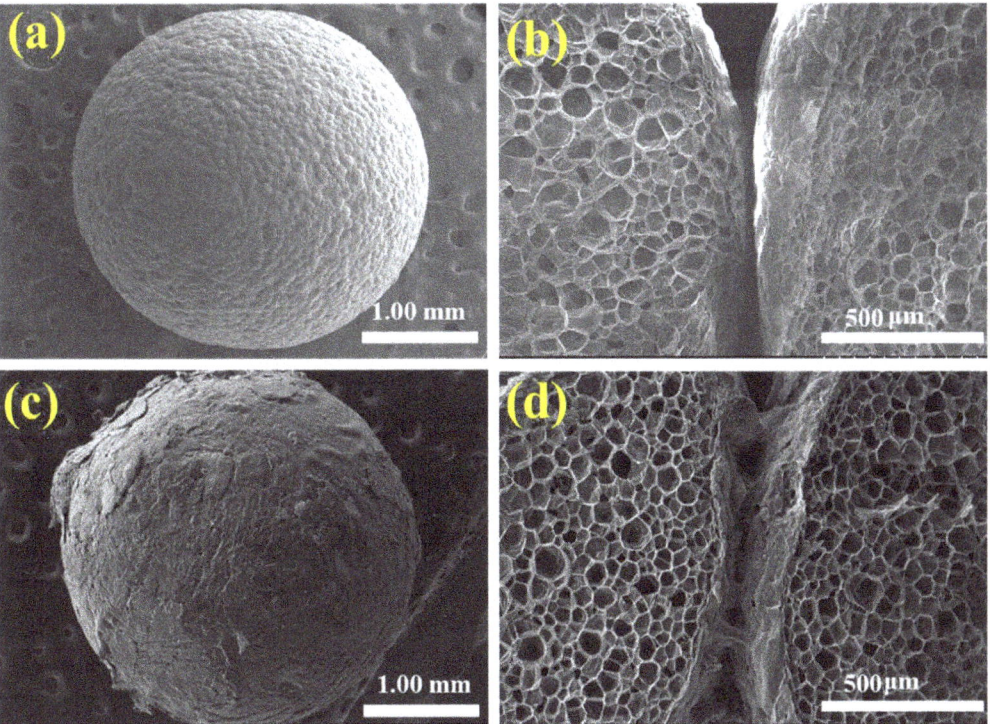

**Figure 2.** The SEM and cross-sectional SEM images of the neat EPS (**a**,**b**), and the EPS/HFR12 (**c**,**d**).

*3.3. Combustion Behavior*

The photographic images of the neat EPS and the various flame-retardant EPS foams that were captured after the combustion test are presented in Figure 3. The neat EPS sample was observed to generate a smoky and sooty flame during the combustion process, and no residue was detected after combustion (Figure 3a). By contrast, the image in Figure 3b reveals the broken and expanded char layer and somewhat collapsed structure of the combusted EPS/HFR0. Moreover, although a similar char residue was observed for the flame-retardant EPS/HFR9 sample shown in Figure 3c, the structural integrity was better preserved than in the EPS/HFR0 sample. These results clearly demonstrate the improved flame-retardant performance of the coated EPS foam. Further, the increased gypsum content in the EPS/HFR12 sample was found to generate a dense and compact char foam without any cracking (Figure 3d). This can provide an even more effective thermal barrier, thus further enhancing the flame retardancy. However, the further increase in gypsum contents for the EPS/HFR15 sample led to the formation of voids in the char layer

(Figure 3e). Here, the compactness and expansion ratio of the char layer were negatively impacted by the large amount of $CaSO_4$, which impeded the diffusion of oxygen, heat, and combustible gases, thereby hindering the decomposition and volatilization of APP. Furthermore, the low integrity of this char layer ultimately reduces its flame-retarding behavior. These results demonstrate that a controlled content of gypsum plays key role in boosting the flame-retardant performance of the EPS foam, with the EPS/HFR12 sample providing the optimum effect.

**Figure 3.** The digital photographs of the combusted samples: (**a**) the neat EPS, (**b**) the EPS/HFR0, (**c**) the EPS/HFR9, (**d**) the EPS/HFR12, and (**e**) the EPS/HFR15.

*3.4. Cone Calorimetry*

The PHRR, THR, and FIGRA curves of the various samples are presented in Figure 4, and the numerical results are summarized in Table 2. Thus, the neat EPS ignited quickly, with high PHRR and THR values of 310.5 kW/m$^2$ and 42.1 MJ/m$^2$, respectively. After coating the hybrid flame-retardant onto the EPS surface, however, the EPS/HFR0 exhibited significantly reduced PHRR and THR values of 67.1 kW/m$^2$ and 15.9 MJ/m$^2$, respectively; these results confirm that the flame-retardant coating acts as a barrier layer during the combustion process. Furthermore, the addition of gypsum was seen to drastically reduce these values to 57.5 kW/m$^2$ and 13.4 MJ/m$^2$, respectively, for the EPS/HFR9 sample, and 53.1 kW/m$^2$ and 8.0 MJ/m$^2$, respectively, for the EPS/HFR12 sample. With the further increase in gypsum content, however, the PHRR and THR values increased slightly to 55.8 kW/m$^2$ and 10.6 MJ/m$^2$, respectively, for the EPS/HFR15 sample. These results further demonstrate that the gypsum-based flame-retardant coating has the potential to improve the flame retardancy of the EPS, with the EPS/HFR12 exhibiting by far the lowest PHRR and THR values.

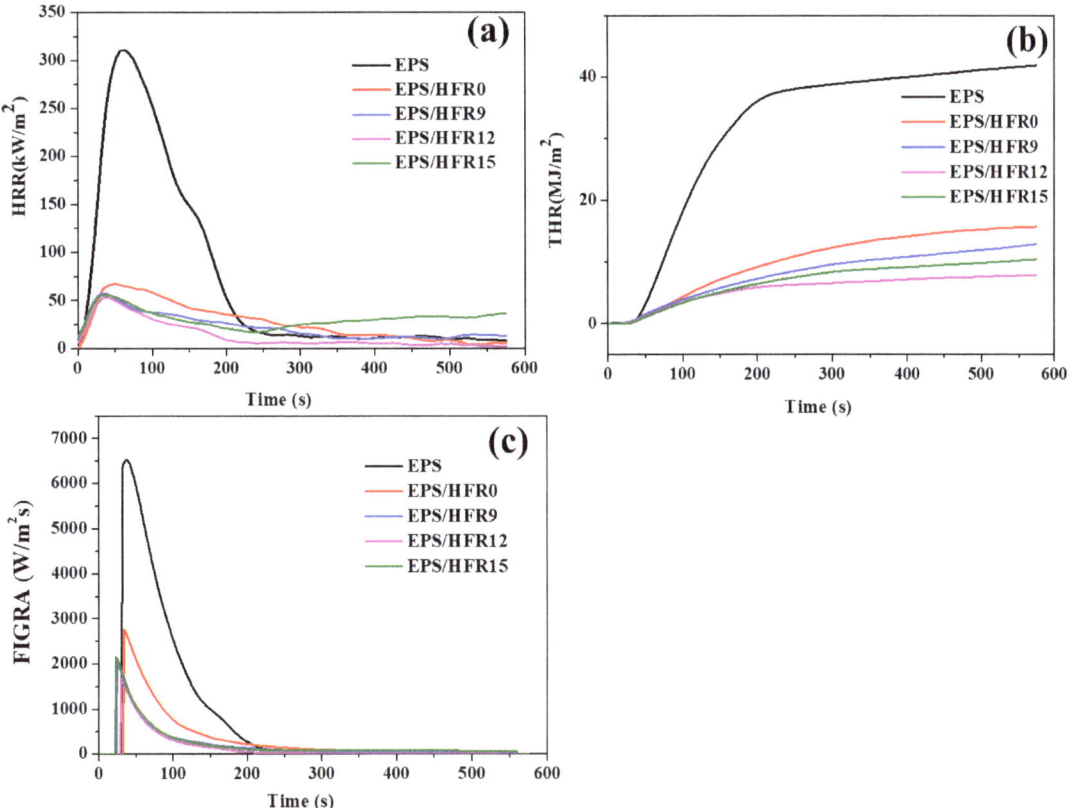

Figure 4. The cone calorimetry curves of the various EPS samples: (**a**) the PHRR curves; (**b**) the THR curves; and (**c**) the FIGRA curves.

Table 2. The cone calorimeter test results for the various EPS foam samples.

| Sample | PHRR (kW/m$^2$) | THR (MJ/m$^2$) | FIGRA (W/m$^2$·s) |
| --- | --- | --- | --- |
| EPS | 310.5 | 42.1 | 6530.8 |
| EPS/HFR0 | 67.1 | 15.9 | 2764.1 |
| EPS/HFR9 | 57.5 | 13.4 | 2119.0 |
| EPS/HFR12 | 53.1 | 8.0 | 1682.9 |
| EPS/HFR15 | 55.8 | 10.6 | 2147.2 |

The underlying mechanism for this improved flame-retarding behavior of the EPS foam in the presence of gypsum is as follows. During the combustion process, the gypsum absorbs the generated heat and releases water molecules to form the thermally stable calcium sulphate. During this endothermic process, a further increase in temperature is delayed until the gypsum is completely dehydrated. The resulting calcium sulphate then provides an effective barrier to further heat flow, thereby reducing the heat transfer during the remainder of the combustion process. The above results therefore demonstrate that gypsum plays a key role in boosting the flame retardancy of the HFR coating during the combustion process. However, the synergistic effect of the gypsum-based HFR relies on a moderate content of gypsum, with higher HRR and THR values observed when the gypsum content is increased in the EPS/HFR15 sample. The excess gypsum releases water to form excess thermally stable $CaSO_4$ on the surface of the char layer, which not only hinders the diffusion of oxygen, heat, and flammable gases, but also hampers the decomposition and volatilization of the APP, thereby hindering the swelling process of the char layer and, thus, reducing the flame-retardant performance.

The burning characteristics of the materials are demonstrated by the FIGRA test results in Figure 4c and Table 2, where the lower FIGRA values of all the flame retardant-based EPS foams relative to the pristine EPS foam indicate the increased fire safety of the composite materials [50,51]. In detail, the EPS, EPS/HFR0, EPS/HFR9, EPS/HFR12, and EPS/HFR15 foams exhibit FIGRA values of 6530.8, 2764.1, 2119.0, 1682.9, and 2147.2 $W/m^2$ s, respectively. Thus, the lowest FIGRA value was obtained with the flame-retardant EPS/HFR12 sample.

*3.5. Char Residue Analysis*

It is well known that the flame-retardant performance of the composite material depends on the compactness of the char layer that is clearly generated during the combustion process [52]. Hence, the role of the gypsum additive in the flame-retardant coating on the EPS foam was further elucidated by the SEM and EDS analysis of the char layer obtained in the CC test in the presence and absence of gypsum (Figure 5). Thus, the surface of the residual char layer on the EPS/HFR0 sample clearly exhibits some collapse structures and holes, which facilitate the release of large amounts of heat during the combustion process (Figure 5a). Further, the EDS analysis in Figure 5b indicates that the char residue on the EPS/HFR0 sample contains only C, O, Si, P, and Ca, with no S. In the presence of gypsum, however, the formation of a dense and compact residual char layer was detected on the EPS/HFR12 sample (Figure 5c), with severely limited voids compared to the EPS/HFR0 sample (Figure 5a). This more compact char would be beneficial for reducing heat transfer during the combustion process. Moreover, the EDX spectrum of the char residue on the EPS/HFR12 sample in Figure 5d reveals the presence of S, along with higher Ca and O contents than in the EPS/HFR0 (Figure 5b), clearly indicating the presence of the thermally stable $CaSO_4$. Thus, the SEM-EDS results confirm the beneficial effects of the gypsum additive in enhancing the flame-retardant performance of the EPS foam not only via the endothermic dehydration process and formation of the insulating $CaSO_4$ layer, as detailed above, but also via the expansion of the EG, which releases non-flammable gases and generates a worm-like char layer. Similarly, the APP content of the binder begins to decompose and release incombustible gases such as $NH_3$ and $H_2O$, thereby resulting in the formation of polyphosphoric acid, as detailed in Section 3.1. The esterification reaction between this phosphoric acid and the hydroxyl group of the PER in the binder then results in the formation of a char framework. In addition, the DBDPE content of the binder decomposes and releases bromine radicals, which accelerate the gas phase and suppress the spreading of the flame. Meanwhile, the talc and $CaCO_3$ components react with the phosphoric network to form a silicon phosphate and calcium phosphate, thereby resulting in the formation of a thermally stable and dense, compact char, which further enhances the flame resistance of the EPS foam by helping to reduce the PHRR and THR during combustion. Thus, the as-fabricated gypsum-based HFR materials can effectively limit the combustion process of the EPS material.

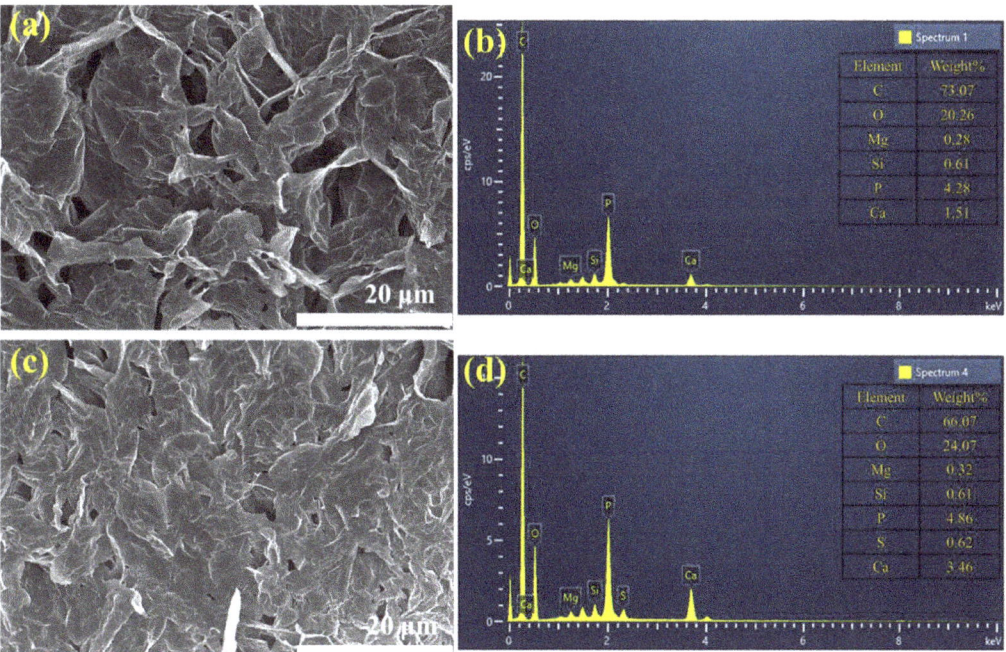

**Figure 5.** The SEM images (**a,c**) and corresponding EDS spectra (**b,d**) of the char residue on the EPS/HFR0 (**a,b**) and the EPS/HFR12 (**c,d**).

*3.6. Physical Properties*

In order to use a material for thermal insulation application, the EPS foam must not only be able to fulfill the demand for flame-retardant performance, but it should also possess essential physical properties, such as density and thermal conductivity. The influence of gypsum with hybrid flame retardant materials on the density and thermal conductivity of EPS foams was tested, and the results are listed in Table 3. In comparison to neat EPS foam, the density of flame-retardant-based EPS foam increased. The neat EPS foam exhibited a density of 26 kg/m$^3$, which increased to 68 kg/m$^3$ for EPS/HFR0. We observe that with the incorporation of gypsum, the density of flame-retardant EPS improved to up to 71 kg/m$^3$ (for EPS/HFR9), 72 kg/m$^3$ (for EPS/HFR12), and 74 kg/m$^3$ (for EPS/HFR15). This increased density may be caused by the uniform adhesion of the flame-retardant coatings on the surface of the EPS beads, which may be attributed to the increase in gypsum content in hybrid flame retardant systems.

**Table 3.** Physical properties of the neat EPS and flame-retardant-based EPS foam.

| Sample | Density (kg/m$^3$) | Thermal Conductivity (W/m.K) |
| --- | --- | --- |
| EPS | 26 | 0.028 |
| EPS/HFR0 | 68 | 0.038 |
| EPS/HFR9 | 71 | 0.038 |
| EPS/HFR12 | 72 | 0.038 |
| EPS/HFR15 | 74 | 0.038 |

The thermal conductivity is a vital index for measuring the thermal insulation performance of EPS foam. This signifies its suitability in thermal insulating applications. Our results indicate that neat EPS exhibits a very low thermal conductivity 0.028 W/m.K. However, we observed an enhanced thermal conductivity of up to 0.038 W/m.K for the

EPS/HFR0 system. It was also found that the thermal conductivity of flame-retardant EPS with and without gypsum contents remained unchanged and was observed as identical to the 0.038 W/m.K value. It was determined that the thermal conductivity of flame-resistant EPS foam may be significantly enhanced with acceptable thermal insulation properties.

## 4. Conclusions

A gypsum-based hybrid flame retardant (HFR) system was prepared herein in order to boost the flame-retardant performance of expanded polystyrene (EPS)-based foam materials. The morphological analysis confirmed that the gypsum-based HFR layer was uniformly coated on the EPS beads. In addition, thermogravimetric analysis (TGA) showed that the gypsum significantly enhanced the final residual weight at 800 °C. Importantly, the cone calorimetry test (CCT) results showed that the addition of an optimum amount (12 g per 55 g of binder) of gypsum effectively reduced the peak heat release rate (PHRR), total heat release (THR), and fire growth rate (FIGRA) values to 53.1 kW/m$^2$, 8 MJ/m$^2$ and 1682.95 W/m$^{-2}$ s, respectively. In addition, the char residue analysis demonstrated that the incorporation of gypsum provides a thermally stable and compact char layer, thereby boosting the flame-retardant properties of the EPS foam. However, an excessive amount of gypsum (15 g per 55 g of binder) was found to restrict the formation of the hybrid char products and destroy the swelling behavior of the charred layer, thereby compromising the flame-retardant performance of the HFR. The authors believe that the addition of the optimum amount of gypsum (12 g per 55 g of binder) provides HFR with promising flame-retardance and satisfies the fire-safety standards.

**Author Contributions:** Conceptualization, C.K.H. and K.-H.P.; Methodology, S.P.B.; Validation, C.K.H., J.K. and W.J.; Formal analysis, S.P.B. and J.K.; Investigation, C.K.H.; Writing—original draft preparation, S.P.B.; Writing—review and editing, S.P.B., J.K., P.H.B. and S.S.M.; Supervision, C.K.H. and K.-H.P. Project administration, C.K.H. All authors have read and agreed to the published version of the manuscript.

**Funding:** This research received no external funding.

**Institutional Review Board Statement:** Not applicable.

**Informed Consent Statement:** Not applicable.

**Data Availability Statement:** All data has been provided in within this manuscript.

**Conflicts of Interest:** The authors declare no conflict of interest.

## References

1. Ji, W.; Yao, Y.; Guo, J.; Fei, B.; Gu, X.; Li, H.; Sun, J.; Zhang, S. Toward an understanding of how red phosphorus and expandable graphite enhance the fire resistance of expandable polystyrene foams. *J. Appl. Polym. Sci.* **2020**, *137*, 49045. [CrossRef]
2. Shao, X.; Du, Y.; Zheng, X.; Wang, J.; Wang, Y.; Zhao, S.; Xin, Z.; Li, L. Reduced fire hazards of expandable polystyrene building materials via intumescent flame-retardant coatings. *J. Mater. Sci.* **2020**, *55*, 7555–7572. [CrossRef]
3. Demirel, B. Optimization of the composite brick composed of expanded polystyrene and pumice blocks. *Constr. Build. Mater.* **2013**, *40*, 306–313. [CrossRef]
4. Raps, D.; Hossieny, N.; Park, C.B.; Altstädt, V. Past and present developments in polymer bead foams and bead foaming technology. *Polymer* **2015**, *56*, 5–19. [CrossRef]
5. Cao, B.; Gu, X.; Song, X.; Jin, X.; Liu, X.; Sun, J.; Zhang, S. The flammability of expandable polystyrene foams coated with melamine modified urea formaldehyde resin. *J. Appl. Polym. Sci.* **2017**, *134*, 44423. [CrossRef]
6. Wang, L.; Wang, C.; Liu, P.; Jing, Z.; Ge, X.; Jiang, Y. The flame resistance properties of expandable polystyrene foams coated with a cheap and effective barrier layer. *Constr. Build. Mater.* **2018**, *176*, 403–414. [CrossRef]
7. Lu, H.; Wilkie, C.A. Study on intumescent flame retarded polystyrene composites with improved flame retardancy. *Polym. Degrad. Stab.* **2010**, *95*, 2388–2395. [CrossRef]
8. Wang, S.; Chen, H.; Liu, N. Ignition of expandable polystyrene foam by a hot particle: An experimental and numerical study. *J. Hazard. Mater.* **2015**, *283*, 536–543. [CrossRef]
9. Antonatus, E. Fire safety of etics with EPS material properties and relevance for fire safety during transport, construction and under end use conditions in external thermal insulation component systems. *MATEC Web Conf.* **2013**, *9*, 02008. [CrossRef]

10. Khanal, S.; Zhang, W.; Ahmed, S.; Ali, M.; Xu, S. Effects of intumescent flame retardant system consisting of tris (2-hydroxyethyl) isocyanurate and ammonium polyphosphate on the flame retardant properties of high-density polyethylene composites. *Compos. Part A Appl. Sci. Manuf.* **2018**, *112*, 444–451. [CrossRef]
11. Bensabath, T.; Sarazin, J.; Jimenez, M.; Samyn, F.; Bourbigot, S. Intumescent polypropylene: Interactions between physical and chemical expansion. *Fire Mater.* **2021**, *45*, 387–395. [CrossRef]
12. Da Silveira, M.R.; Peres, R.S.; Moritz, V.F.; Ferreira, C.A. Intumescent coatings based on tannins for fire protection. *Mater. Res.* **2019**, *22*, e20180433. [CrossRef]
13. Maqsood, M.; Langensiepen, F.; Seide, G. The efficiency of biobased carbonization agent and intumescent flame retardant on flame retardancy of biopolymer composites and investigation of their melt-spinnability. *Molecules* **2019**, *24*, 1513. [CrossRef] [PubMed]
14. Qi, F.; Tang, M.; Wang, N.; Liu, N.; Chen, X.; Zhang, Z.; Zhang, K.; Lu, X. Efficient organic-inorganic intumescent interfacial flame retardants to prepare flame retarded polypropylene with excellent performance. *RSC Adv.* **2017**, *7*, 31696–31706. [CrossRef]
15. Li, X.L.; Zhang, F.H.; Jian, R.K.; Ai, Y.F.; Ma, J.L.; Hui, G.J.; Wang, D.Y. Influence of eco-friendly calcium gluconate on the intumescent flame-retardant epoxy resin: Flame retardancy, smoke suppression and mechanical properties. *Compos. Part B Eng.* **2019**, *176*, 107200. [CrossRef]
16. Bhoite, S.P.; Kim, J.; Jo, W.; Bhoite, P.H.; Mali, S.S.; Park, K.H.; Hong, C.K. Expanded polystyrene beads coated with intumescent flame retardant material to achieve fire safety standards. *Polymers* **2021**, *13*, 2662. [CrossRef]
17. Beh, J.H.; Yew, M.C.; Saw, L.H.; Yew, M.K. Fire resistance and mechanical properties of intumescent coating using novel bioash for steel. *Coatings* **2020**, *10*, 1117. [CrossRef]
18. Li, L.; Shao, X.; Zhao, Z.; Liu, X.; Jiang, L.; Huang, K.; Zhao, S. Synergistic Fire Hazard Effect of a Multifunctional Flame Retardant in Building Insulation Expandable Polystyrene through a Simple Surface-Coating Method. *ACS Omega* **2020**, *5*, 799–807. [CrossRef]
19. Choi, J.; Lee, G.; Kim, S.; Choi, K. Investigation on sex hormone-disruption effects of two novel brominated flame retardants (Dbdpe and btbpe) in male zebrafish (danio rerio) and two human cell lines (h295r and mvln). *Appl. Sci.* **2021**, *11*, 3837. [CrossRef]
20. Yu, G.; Bu, Q.; Cao, Z.; Du, X.; Xia, J.; Wu, M.; Huang, J. Brominated flame retardants (BFRs): A review on environmental contamination in China. *Chemosphere* **2016**, *150*, 479–490. [CrossRef]
21. Covaci, A.; Harrad, S.; Abdallah, M.A.E.; Ali, N.; Law, R.J.; Herzke, D.; de Wit, C.A. Novel brominated flame retardants: A review of their analysis, environmental fate and behaviour. *Environ. Int.* **2011**, *37*, 532–556. [CrossRef] [PubMed]
22. Yang, Z.; Peng, H.; Wang, W.; Liu, T. Crystallization behavior of poly(ε-caprolactone)/layered double hydroxide nanocomposites. *J. Appl. Polym. Sci.* **2010**, *116*, 2658–2667. [CrossRef]
23. Wang, J.; Chen, S.; Nie, X.; Tian, M.; Luo, X.; An, T.; Mai, B. Photolytic degradation of decabromodiphenyl ethane (DBDPE). *Chemosphere* **2012**, *89*, 844–849. [CrossRef] [PubMed]
24. Huhtala, S. In vivo and vitro toxicity of decabromodiphenyl ethane, a flame retardant. *Environ. Toxicol.* **2009**, *25*, 333–338. [CrossRef]
25. Hornsby, P.R. Fire retardant fillers for polymers. *Int. Mater. Rev.* **2001**, *46*, 199–210. [CrossRef]
26. Braun, U.; Schartel, B. Flame retardant mechanisms of red phosphorus and magnesium hydroxide in high impact polystyrene. *Macromol. Chem. Phys.* **2004**, *205*, 2185–2196. [CrossRef]
27. Hull, T.R.; Witkowski, A.; Hollingbery, L. Fire retardant action of mineral fillers. *Polym. Degrad. Stab.* **2011**, *96*, 1462–1469. [CrossRef]
28. Chen, W.; Jiang, Y.; Qiu, R.; Xu, W.; Hou, Y. Investigation of UiO-66 as Flame Retardant and Its Application in Improving Fire Safety of Polystyrene. *Macromol. Res.* **2020**, *28*, 42–50. [CrossRef]
29. Alongi, J.; Han, Z.; Bourbigot, S. Intumescence: Tradition versus novelty. A comprehensive review. *Prog. Polym. Sci.* **2014**, *51*, 28–73. [CrossRef]
30. Rajaei, M.; Wang, D.Y.; Bhattacharyya, D. Combined effects of ammonium polyphosphate and talc on the fire and mechanical properties of epoxy/glass fabric composites. *Compos. Part B Eng.* **2017**, *113*, 381–390. [CrossRef]
31. Xu, Z.; Chu, Z.; Yan, L.; Chen, H.; Jia, H.; Tang, W. Effect of chicken eggshell on the flame-retardant and smoke suppression properties of an epoxy-based traditional APP-PER-MEL system. *Polym. Compos.* **2019**, *40*, 2712–2723. [CrossRef]
32. Pedersen, B.F.; Semmingsen, D. Neutron Diffraction Refinement of the Structure of Gypsum, $CaSO_4 \cdot 2H_2O$. *Acta Crystallogr. Sect. B Struct. Crystallogr. Cryst. Chem.* **1982**, *38*, 1074–1077.
33. Thomas, G. Thermal properties of gypsum plasterboard at high temperatures. *Fire Mater.* **2002**, *26*, 37–45. [CrossRef]
34. Charola, A.E.; Pühringer, J.; Steiger, M. Gypsum: A review of its role in the deterioration of building materials. *Environ. Geol.* **2007**, *52*, 207–220. [CrossRef]
35. Javangula, H.; Lineberry, Q. Comparative studies on fire-rated and standard gypsum wallboard. *J. Therm. Anal. Calorim.* **2014**, *116*, 1417–1433. [CrossRef]
36. Ballirano, P.; Melis, E. Thermal behaviour and kinetics of dehydration of gypsum in air from in situ real-time laboratory parallel-beam X-ray powder diffraction. *Phys. Chem. Miner.* **2009**, *36*, 391–402. [CrossRef]
37. Borreguero, A.M.; Luz Sánchez, M.; Valverde, J.L.; Carmona, M.; Rodríguez, J.F. Thermal testing and numerical simulation of gypsum wallboards incorporated with different PCMs content. *Appl. Energy* **2011**, *88*, 930–937. [CrossRef]

38. Puri, R.G.; Khanna, A.S. Effect of cenospheres on the char formation and fire protective performance of water-based intumescent coatings on structural steel. *Prog. Org. Coatings* **2016**, *92*, 8–15. [CrossRef]
39. Kannan, P.; Biernacki, J.J.; Visco, D.P.; Lambert, W. Kinetics of thermal decomposition of expandable polystyrene in different gaseous environments. *J. Anal. Appl. Pyrolysis* **2009**, *84*, 139–144. [CrossRef]
40. Chiang, C.L.; Hsu, S.W. Novel epoxy/expandable graphite halogen-free flame retardant composites-preparation, characterization, and properties. *J. Polym. Res.* **2010**, *17*, 315–323. [CrossRef]
41. Duquesne, S.; Le Bras, M.; Bourbigot, S.; Delobel, R.; Camino, G.; Eling, B.; Lindsay, C.; Roels, T. Thermal degradation of polyurethane and polyurethane/expandable graphite coatings. *Polym. Degrad. Stab.* **2001**, *74*, 493–499. [CrossRef]
42. Gao, S.; Zhao, X.; Liu, G. Synthesis of an integrated intumescent flame retardant and its flame retardancy properties for polypropylene. *Polym. Degrad. Stab.* **2017**, *138*, 106–114. [CrossRef]
43. Guo, J.; Wang, M.; Li, L.; Wang, J.; He, W.; Chen, X. Effects of thermal-oxidative aging on the flammability, thermal degradation kinetics and mechanical properties of DBDPE flame retardant long glass fiber reinforced polypropylene composites. *Polym. Compos.* **2018**, *39*, E1733–E1741. [CrossRef]
44. Chen, X.S.; Yu, Z.Z.; Liu, W.; Zhang, S. Synergistic effect of decabromodiphenyl ethane and montmorillonite on flame retardancy of polypropylene. *Polym. Degrad. Stab.* **2009**, *94*, 1520–1525. [CrossRef]
45. Luo, X.; He, M.; Guo, J.B.; Wu, B. Flame retardancy and mechanical properties of brominated flame retardant for long glass fiber reinforced polypropylene composites. *Adv. Mater. Res.* **2013**, *750–752*, 85–89. [CrossRef]
46. Deodhar, S.; Shanmuganathan, K.; Fan, Q.; Wilkie, C.A.; Costache, M.C.; Dembsey, N.A.; Patra, P.K. Calcium carbonate and ammonium polyphosphate-based flame retardant composition for polypropylene. *J. Appl. Polym. Sci.* **2011**, *120*, 1866–1873. [CrossRef]
47. Wang, F.; Liu, H.; Yan, L. Comparative study of fire resistance and char formation of intumescent fire-retardant coatings reinforced with three types of shell bio-fillers. *Polymers* **2021**, *13*, 4333. [CrossRef]
48. Singh, K.; Ohlan, A.; Saini, P.; Dhawan, S.K. composite—Super paramagnetic behavior and variable range hopping 1D conduction mechanism—Synthesis and characterization. *Polym. Adv. Technol.* **2008**, *19*, 229–236. [CrossRef]
49. Ghazi Wakili, K.; Hugi, E.; Wullschleger, L.; Frank, T. Gypsum board in fire—Modeling and experimental validation. *J. Fire Sci.* **2007**, *25*, 267–282. [CrossRef]
50. Tsai, K.C. Orientation effect on cone calorimeter test results to assess fire hazard of materials. *J. Hazard. Mater.* **2009**, *172*, 763–772. [CrossRef]
51. Seo, D.; Kim, D.; Kim, B.; Kwon, Y. An experimental study on the combustibles investigation and fire growth rate for predicting initial fire behavior in building. *Procedia Eng.* **2013**, *62*, 671–679. [CrossRef]
52. Zhang, W.; Wu, W.; Meng, W.; Xie, W.; Cui, Y.; Xu, J.; Qu, H. Core-shell graphitic carbon nitride/zinc phytate as a novel efficient flame retardant for fire safety and smoke suppression in epoxy resin. *Polymers* **2020**, *12*, 212. [CrossRef] [PubMed]

*Review*

# Interfacial Adhesion in Silica-Silane Filled NR Composites: A Short Review

Chang Seok Ryu [1] and Kwang-Jea Kim [2],*

[1] School of Chemical Engineering, Chonnam National University, Gwangju 61186, Korea; netryu@hanmail.net
[2] DTR VMS Italy S.r.l., 25050 Brescia, Italy
* Correspondence: kwangjea.kim@gmail.com

**Abstract:** We reviewed the accelerators, the hydrolysis and condensation reaction mechanism of bifunctional alkoxy silane, and the mechanism of zinc ion in natural rubber (NR) composites. NR composites transform into thermoset composites after vulcanization reaction with help of sulfur and accelerators. Bifunctional alkoxy silanes chemically bond between NR and inorganic silica. For alkoxy silane coupling with silica surface, hydrolysis reaction takes first and then condensation reaction with hydroxyl group in silica takes place. With help of zinc ion the reaction efficiency increases significantly. Zinc ion, a smart material that increases accelerator synergy, mechanism for improvements of interfacial adhesion between NR and silica was revisited.

**Keywords:** silica; silane; natural rubber (NR); hydrolysis; interfacial adhesion; zinc mechanism

## 1. Introduction

Silica-silane filled natural rubber (NR) composites, as a tire tread material, are currently used in various fields as a smart composite replacing the conventional carbon black (CB) tire composites. Compared with CB filled NR tire, it shows excellent performance such as improved fuel efficiency due to low rolling resistance, reduced braking distance due to slip resistance due to high tan delta, and improved traction in snow and wet conditions when driving a vehicle. Currently many vehicles use a green tire (silica filled tire) [1]. It is called green tire because it has green strength during silica composite processing. The green tire composite has differences in interfacial bonding compared to the CB compound. And even now, various efforts are being made to increase the efficiency of interfacial bonding. Compared with the CB filled NR composite system, in both systems the interfacial interaction in the silica-silane filled NR composite has the same mechanism for chemical bonding of rubber by sulfur, but the interfacial interaction between filler and NR is different. That is, the CB filled system forms a bond by interaction between CB and NR by increasing compatibility due to the π-π interaction between the double bonds of the two materials. However, in the silica filled system, as bifunctional silane chemically bonds silica and NR, the interfacial interaction between the two materials increases by chemical bonding.

Until now, there was no published article introducing the mechanism for the entire process of the chemical reaction from the raw material to the silica-silane-NR composite to increase the interfacial interaction. This paper briefly summarizes the whole process above. We first reviewed the accelerators that affect the degree and rate of bonding between rubber chains, and reviewed the hydrolysis mechanism of alkoxy silane and the condensation reaction mechanism between hydrolyzed silane and hydroxyl group on silica surface. And the chemical reaction mechanism between bifunctional silane and rubber was reviewed. Zinc ion, a smart material that effectively increases accelerator synergy and increases the effectiveness of vulcanization, was also reviewed.

**Citation:** Ryu, C.S.; Kim, K.-J. Interfacial Adhesion in Silica-Silane Filled NR Composites: A Short Review. *Polymers* 2022, 14, 2705. https://doi.org/10.3390/polym14132705

Academic Editor: Vincenzo Fiore

Received: 15 April 2022
Accepted: 15 June 2022
Published: 1 July 2022

**Publisher's Note:** MDPI stays neutral with regard to jurisdictional claims in published maps and institutional affiliations.

**Copyright:** © 2022 by the authors. Licensee MDPI, Basel, Switzerland. This article is an open access article distributed under the terms and conditions of the Creative Commons Attribution (CC BY) license (https://creativecommons.org/licenses/by/4.0/).

## 2. Materials and Methods

### 2.1. Accelerators

Various additives are added to rubber composite such as filler, plasticizer, compatibilizer, anti-degradant, processing aid, curative, accelerator, etc. A typical curative used in rubber is sulfur, which chemically promotes bonding between rubber chains to increase interfacial adhesion between chains. Accelerators play a role in controlling the degree and rate of chemical reaction between sulfur and rubber chains. Cure time or degree of crosslinking is different depending on the type of accelerator thus the choice of accelerator is different depending on the composite and processing purpose. First, we revisit the history of accelerators.

Goodyear-Hancock first showed unaccelerated vulcanization of rubber [2,3]. Later, organic accelerator, aniline, was first introduced in 1906 [4]. However, aniline was too toxic for use in rubber vulcanization. Later, reaction product of aniline with carbon disulfide such as thiocarbanilide was introduced in 1907 [5]. Subsequently carbon disulfide modification with aliphatic amines (e.g., dithiocarbamates), which thiuram type accelerators were introduced in 1919 [6]. And then delayedaction accelerators such as 2-mercaptobenzothiazole (MBT) and 2-mercaptobenzothiazole disulfide (or 2,2'-dithiobisbenzothiazole) (MBTS) were introduced in 1925 [7–9]. The first commercial benzothiazole sulfenamide (N-cyclohexylbenzothiazole 2-sulfenamide), a delayed-action accelerator, was introduced by Harman in 1937 [10]. More delayed-action premature vulcanization inhibitor (PVI), N-(cyclohexylthio)phthalimide (CTP), was introduced in 1968 [11]. Studies on accelerator effects on vulcanization were focused mainly on CB filled systems. Various types of accelerators and their roles in CB filled rubber compounds were summarized [6]. Thurn et al. [12] first discovered the use of silica in combination with bis(3-triethoxysilylpropyl) tetrasulfane (TESPT) and NR, and this was patented for practical 'green tire' application in 1992 [13]. Since then, considerable efforts have been made on studies of silica-silane reinforcement in rubber compounds [14], silica dispersion [15–19], silica-silane reaction [15,16,20–28], silane-rubber reaction [29–31], zinc ion- [26,32–35], moisture- [36,37], temperature- [38], and accelerator effects on vulcanization properties [39,40], effects of processing geometry [19], and effects of polymer blends [27,28,41,42].

There were many studies on accelerator effects on vulcanization properties of CB filled rubber compounds [39,40,43–45]. For example, thiazole type accelerators on crosslinking rate in the NR compound [43] and NR & synthetic poly-1,5-dienes [44] were studied. Sulfenamide type accelerators on rate of sulfenamide-sulfur reaction [45] and scorch delay reaction in the NR & SBR compound [46] were studied. Vulcanization rate of silica/NR compounds and CB/NR compounds, which contain various accelerators were directly compared [1,35,40]. These accelerators consist of different chemical structure, i.e., thiuram type, TMTD (tetramethylthiuram disulfide) and DPTT (dipentamethylenethiuram tetrasulfide); thiazole type, MBT (2-mercaptobenzothiazole) and MBTS (2,2dithiobis(benzothiazole)); sulfenamide type, CBS (n-cyclohexylbenzothiazyl-2-sulfenamide) and NOBS (n-oxydiethylenebenzothiazyl-2-sulfenamide)]; and zinc ion containing thiuram type ZDBC (zinc di-n-butyldithiocarbamate)]. The order of accelerator reaction was thiuram(TMTD,DPTT), thiazole(MBT,MBTS), and then sulfonamide(CBS,NOBS) types due to their chemical structure [1].

### 2.2. Silica vs. CB

Comparing silica surface with CB, silica surface consists of Si-O bonds with hydroxyl groups (-OH), while CB mainly consists of C-C and C=C bonds. Silica particles exhibit a strong filler-filler interaction due to its polar ($Si^{+\delta}$-$O^{-\delta}$) character due to differences in electro negativity between Si(3.5) and O(1.9) thus easily agglomerate each other, while CB shows a weak C=C interaction due to the presence of a π-π bond, thereby showing good compatibility with rubber chains, which contain a double bond. The π-π bonds and/or oxygen on the CB surface interact with double bonds (π-π) in the rubber chain, which shows good compatibility with the rubber chain. Therefore, the dispersion of CB is better than that of silica in the rubber matrix. On the other hand, modification of the silica surface with silanes, particularly with bifunctional organosilanes, such as bis(triethoxysilyilpropyl)disulfide (TESPT) and bis(triethoxysilyilpropyl)tetrasulfide (TESPD), assists in the dispersion of silica agglomerates in the rubber chain [18,24]. The chemical reaction between silica and silane as well as silane and rubber modifies the network of the rubber matrix [23]. It has been known that hydroxyl group on silica surface chemically reacts with hydroxyl group in hydrolyzed silane via condensation reaction, and then forms a 3-dimensional (3D) network structure with a rubber chain [15,16,23]. Hence, the filler network constant (filler network interaction parameter) of silane-modified silica is different from CB filled systems. This makes an interpretation of the in-rubber structure of the modified silica system more difficult.

In silica/NR composites, the order of accelerator reaction was the same as shown in the CB filled system, i.e., thiuram (TMTD, DPTT), thiazole (MBT, MBTS), and then sulfonamide (CBS, NOBS) types due to their chemical structure. However, silica/NR composites showed a slower vulcanization rate (ts12, t10, t90) than the CB/NR ones for each accelerator. This is due to two stage reaction in silica-silane/NR composite, which are hydrolysis of alkoxy silane and condensation reaction between hydrolyzed silane and silica surface [1].

*2.3. Interfacial Interaction*

For, interfacial interaction, Wolff and his coworkers introduced the in-rubber filler structure ($\alpha F$) for quantitative measurements of the filler structure [47–51]. Wolff et al. [50] reported that the $\alpha F$ was constant over the entire range of carbon black filled in NR and NBR systems. Their research focused mainly on the filler-filler interaction in rubber compounds. On the other hand, in the case of a silica filled system, $\alpha F$ increased with increasing filler concentration. Kim reviewed his research on the in-rubber structure of CB and silane-treated silica [52].

*2.4. Hydrolysis Mechanism*

The hydrolysis mechanisms as well as the hydrolysis measurement technique and its practical applications in manufacturing fields are revised by Kim [53,54]. This section reviews the theoretical aspects of acid-catalyzed and base-catalyzed hydrolysis mechanisms.

2.4.1. Acid Catalyzed Hydrolysis

Hydrolysis means that moisture is applied to molecular chains, causing the chains of molecules affected by moisture to break down. Moisture present in the composite affects hydrolysis. When in contact with high-purity moisture in a neutral, low ionic state in a non-glass container, silane maintains a stable state for several weeks or months. However, when in contact with tap water, the hydrolysis of alkoxysilane proceeds considerably within a few hours [55]. The catalyst applied to the hydrolysis of alkoxysilane is also applied to the progress of the condensation reaction, and the waterborne silane improves adhesion to the glass surface than the control silane [55]. This means that the catalyst present in moisture significantly increases the silane bond on the silica surface. The primary reaction increases the silanization reaction in the presence of moisture, and the secondary reaction increases with higher moisture content [56–58]. Moisture on the silica surface favors the reaction between the silica and the coupling agent (e.g., TESPD, TESPT), this also reducing the interaction between silica particles [57]. The hydrolysis reaction mechanism also applies for plastics (e.g., polyamide) [59]. Figure 1a,b shows the hydrolysis mechanism by catalysis of base- and acid- of alkoxysilane. The reaction coordinate has a structure of polar parameter values ($\rho^*$) and steric parameter values (s) from the reactant to the transition state. As the reaction proceeds, it qualitatively shows changes and charge distribution [60]. Figure 1a shows the mechanism of acid-catalyzed hydrolysis. Five substituents exist around the silicon atom in the form of SN2-Si. It forms a pentacoordinate intermediate state. The approaching nucleophile and the outgoing leaving group hybridize with the silicon atom, resulting in sp3d hybridization between the nucleophile and the remaining group. The meaning of small values of $\rho^*$ and s is shown in Figure 1a. The reaction mechanism is that after rapid equilibrium protonation in the equilibrium state of the substrate, the water molecule becomes a leaving group and a bimolecular molecule, resulting in SN2-type displacement [61,62].

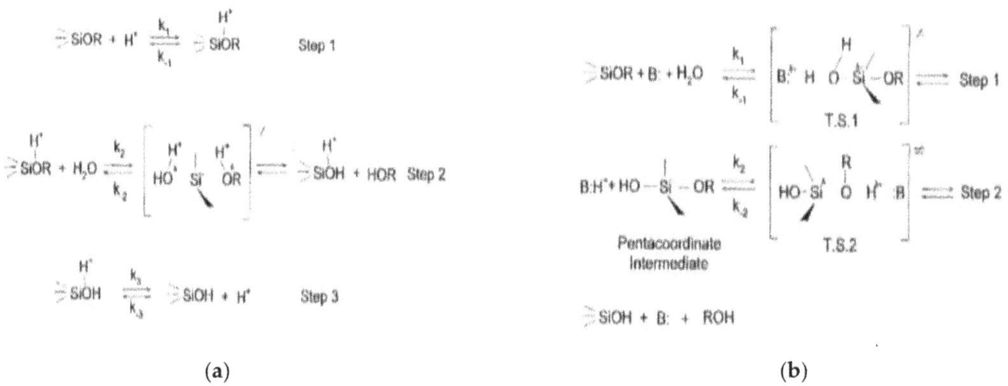

**Figure 1.** Hydrolysis mechanism for (**a**) acid catalyzed (adopted from [61,62]), (**b**) base catalyzed (adopted from [61]) systems.

2.4.2. Base Catalyzed Hydrolysis

The base-catalyzed hydrolysis mechanism is illustrated in Figure 1(b). Bi-molecular nucleophilic substitution reaction (SN2**-Si or SN2*-Si) with the coordinate intermediate; SN2**-Si refers to the rate determination of the pentacoordinate intermediate formation (k1 < k2), and SN2*-Si refers to the rate determination of the break-down (k1 > k2) [61]. A large value of ρ* means that the transition state (T.S.) is more stable than the reactants by electron withdrawing groups attached to silicon atoms. In the proposed mechanism, the negative charge at the silicon atom in the transition state (T.S.1 or T.S.2) is quite high. A large steric parameter (s) means that the transition state is sterically more crowded than the reactant [62,63]. The sp3 hybridized silicon forms bonds with the introduced hydroxide anions during rehybridization, forming the same transition state as sp3d. It forms a strongly coupled transition state between the negative charge density and the nucleophilic oxygen on the silicon surface [64]. Figure 2 shows the hydrolysis mechanism according to the pH of alkoxysilane [65]. As stated by Allen model [66] and Kay & Assink model [67], bas-catalyzed hydrolysis reaction is a condensation reaction when an alkoxy group (-(OR)3) is replaced with a -(OSi)3 group (bottom left); When an alkoxy group (-(OR)3) is substituted with a hydroxy group (-(OH)3) (bottom right), this is called hydrolysis. Acid hydrolysis is significantly greater than base hydrolysis and is minimally affected by other carbon-bonded substituents. The results of Osterholtz & Pohl [68], illustrated by Moore [69], show that when the pH is neutral, hydrolysis occurs last, and it changes 10 times for every 1 pH change. The condensation reaction is also affected by pH, and the minimum reaction pH is 4. The hydrolysis reaction of alcohol is reversible and stabilizes the solution for a long time.

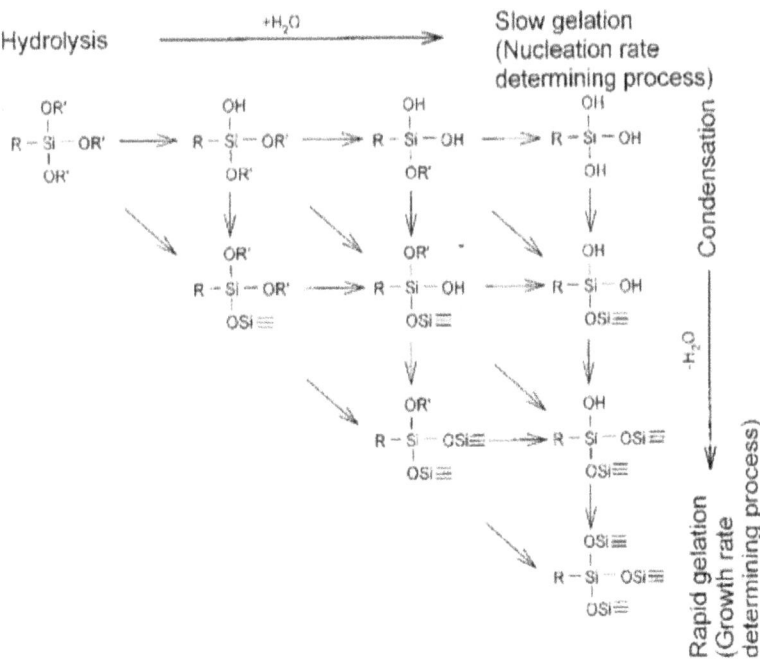

**Figure 2.** Schematic presentation of hydrolysis and condensation reaction of a trialkoxy silane; redraw from [70].

*2.5. Alkoxysilane Silanization Reaction Mechanism*

Bifunctional alkoxysilanes are either direct or go through two reaction steps. Figure 2 shows the first step. After hydrolysis of the alkoxysilane by water molecules, the hydrolyzed silane causes a condensation reaction with the hydroxy group on the silica surface and chemically bonds to the silica surface [70]. In the second step, the sulfur at the other end of the silane chemically reacts with the double bond of the rubber chain. In this step, the sulfur (S8) ring is opened and vulcanized together with the double bond of the rubber chain. Then, a three-dimensional network is formed between silica and rubber [22,37,38,71]. This shows that the higher the degree of hydrolysis, the higher the % of the condensation reaction.

The compatibilizer controls the phase structure by increasing the interfacial adhesion property of the two materials by minimizing the difference in the value of the Flory-Huggins interaction parameter of the two materials when mixing different polymers, and thus, the applied material is reliable. In order to increase the efficiency of interfacial bonding between the resin and the filler, an interfacial adhesive (coupling agent) is treated on the surface of the filler to enhance the interfacial adhesion with the resin to induce chemical bonding at the interface. The roles of the compatibilizer are reinforces the strengths of the two materials (polymer-polymer, or polymer-filler), compensates for the weaknesses, and serves to uniformly disperse the materials [72]. Compatibilizers are mostly classified into non-reactive compatibilizers and reactive compatibilizers. Bifunctional silane is a reactive compatibilizer, which reacts with the double bond between the surface of silica and the rubber chain to chemically bond. Table 1 summarizes the structure, vulcanization form, and application fields of silanes [73].

**Table 1.** Types of silane structure, curing system and application.

| Silane | Structure | Curing System | Application |
|---|---|---|---|
| TESPT | $(C_2H_5O)_3Si-(CH_2)_3-S_4-(CH_2)_3-Si(OC_2H_5)$ | Sulfur | Tire treads, shoe soles, industrial rubber goods |
| TESPD | $(C_2H_5O)_3Si-(CH_2)_3-S_2-(CH_2)_3-Si(OC_2H_5)$ | Sulfur | Tire treads, industrial rubber goods |
| TCPTEO | $(C_2H_5O)_3Si-(CH_2)_3-SCN$ | Sulfur | Shoe soles, industrial rubber goods |
| MTMO | $(CH_3O)_3Si-(CH_2)_3-SH$ | Sulfur | Shoe soles, industrial rubber goods |
| VTEO | $(C_2H_5O)_3Si-CH=CH_2$ | Peroxide | Industrial rubber goods |
| VTEO | $(CH_3-O-C_2H_4O)Si-CH=CH_2$ | Peroxide | Industrial rubber goods |
| CPTEO | $(CH_3O)_3Si-(CH_2)_3-Cl$ | Metal oxide | Chloroprene rubber |
| MEMO | $(CH_3O)_3Si-(CH_2)_3-O-C(O)C(CH_3)=CH_2$ | Peroxide | Textile adhesion |
| AMEO | $(C_2H_5O)_3Si-(CH_2)_3-NH_2$ | Sulfur | Special polymers, metal adhesion |
| OCTEO | $(CH_3-CH_2-O)_3Si-(CH_2)_7-CH_3$ | Sulfur, peroxide | Processing aid |

Figure 3 shows the condensation reaction intermediate between the hydroxyl on the silica surface of a filler (e.g., silica, soft clay) and the hydroxyl of an alkoxy group substituted after hydrolysis of a bifunctional silane (e.g., TESPT) [24,30]. Figure 3 shows that after the condensation reaction, moisture is produced as a by-product. In the production site, the generation of alcohol and moisture during green composite production is regarded as indirect evidence that the material is chemically reacting properly.

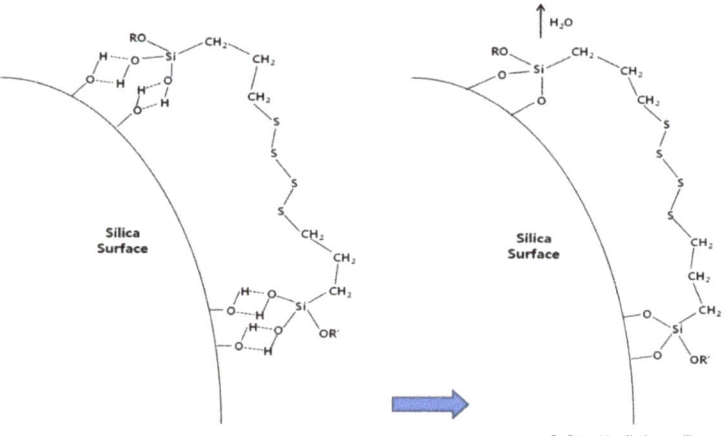

**Figure 3.** Condensation reaction intermediate of silica and hydrolyzed TESPT; redraw from [24,30].

### 3. Hydrolysis Quantitative Analysis

Figure 4 shows the amount of residual alcohol that did not react in TESPD during hydrolysis stage in the composite material measured by the headspace/gas chromatographic(GC) technique [70]. The residual alcohol amount of the composite with water added to silica (open circle) before mixing was lower than that of the composite without water added (closed circle), and as the mixing time increased, the alcohol content of the residual alcohol of both composites decreased. This means that the hydrolysis reaction of silane and silica increases as the presence of moisture increases and the mixing time increases.

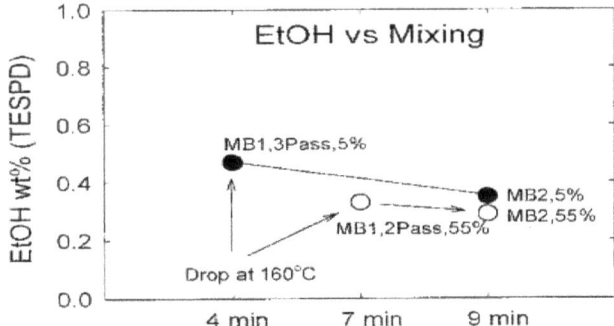

**Figure 4.** Percent of ethoxy remaining in the 3-pass vs. 2-pass compounds; adopted from [70].

Figure 5 shows that the sulfur in the bifunctional silane (e.g., TESPD or TESPT) chain is separated from the bond between the sulfur chains, and the sulfur group at the end forms a covalent bond with the double bond (e.g., NR, SBR) of the rubber chain [42]. Sulfur is further activated with the help of zinc as a reaction activator [74]. That is, when silane is not added, the size of the interaction between filler-polymer(silica-rubber) ($\alpha$FP) is lower than that of CB and rubber chains. However, when silane is added, a covalent bond is created between the silica surface and the rubber chain through the silane as a medium, and the bonding strength is high, so the size of the interaction ($\alpha$FP: filler-polymer interaction parameter in the material) is larger than that of the CB-rubber chain.

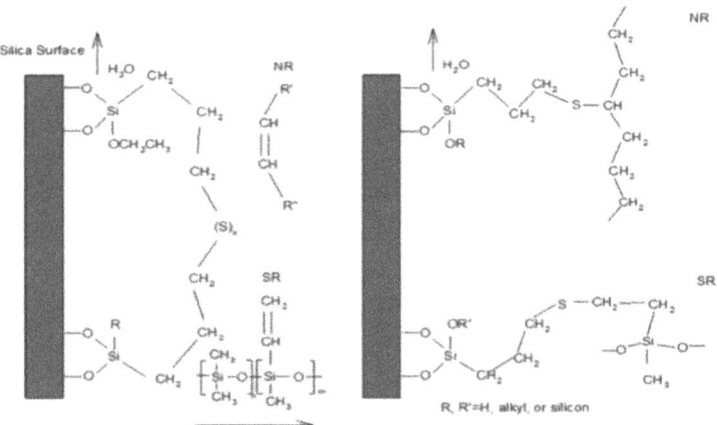

**Figure 5.** Cross-linking of unsaturated elastomers with sulfur containing silane ($x$ = 2 or 4); adopted from [42].

## 4. Effect of Sulfur Concentration of Hydrolyzed Silane

Figure 6 schematically shows the types of bonding between silica and natural rubber (NR) after vulcanization of various types of silanes (e.g., triethoxysilyilpropane (TESP), bis(triethoxysilyilpropyl)disulfide(TESPT), bis(triethoxysilyilpropyl)tetrasulfide (TESPD)) [23]. After silica and silane are combined by a hydrolysis mechanism, ethanol is produced as a by-product and evaporates. Later, the sulfur attacks the double bond present in the NR chain to form a covalent bond with the rubber chain. After vulcanization, it forms a 3D structure with a different structure from silica and NR depending on the type of silane used. Since TESP has no element to react with the double bond of rubber in the molecule, it exists in a state sandwiched between rubber molecules as shown in Figure 6a. In TESPD, sulfur atom attacks the double bond in the NR chain to form a 3D structure as shown in Figure 6b. TESPT also forms a 3D structure as shown in Figure 6c with NR with the same mechanism as TESPD. Since TESPT has an average of 2 more sulfur per molecule compared to TESPD, it has a higher cross-link structure than TESPD after vulcanization.

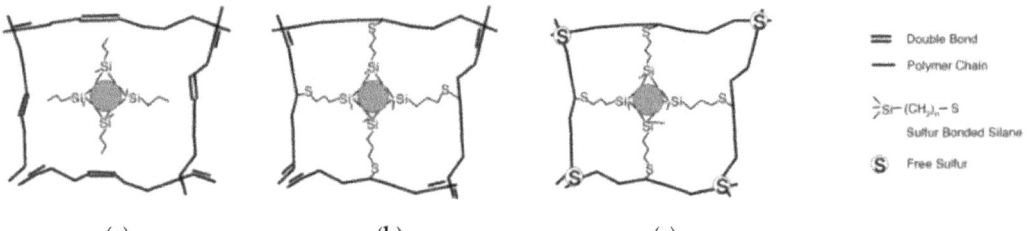

**Figure 6.** Schematic illustration of vulcanized silane treated silica (**a**) TESP (no sulfur), (**b**) TESPD, and (**c**) TESPT; adopted from [23].

Comparing the physical properties of the silica/NR composite treated with TESPT (S4) and treated with TESPD (S2), S4 has a higher tensile modulus and lower maximum tensile stress, and has low tear resistance than S2. At room temperature or 100 °C before and after vulcanization, S4 shows a low tan δ (delta) (=E″/E′) value due to the low E″ (loss modulus). In addition, it can be observed that the tan δ value of both composite decreases as the temperature increases. When the blow out (BO) test was performed on the two vulcanized composites, the S4 composite showed lower recovery from deformation compared to the S2 one. In the HBU (heat build-up) test, S4 composite showed lower heat generation than S2. In the abrasion loss test, S4 composite showed higher loss than S2, which means that S4 has a more developed 3D network structure. However, in some of the rigid 3D structures, the S4 structure may be broken earlier than the S2 when the over critical strain is surpassed.

The results of the BO test of the silica composite treated with silane and without treatment show a lot of difference [75].

Figure 7a shows the photos of the specimen before and after the 'Firestone Flexometer' BO test. Figure 7b shows the deformation ratio (%) of the picture above (Figure 7a). After the BO test compared to the test specimen (No. 0), the NR specimens with only silica (1&2) shows about 6-8% deformation, but the specimens with silica-silane added (3&4 (TESPT(S4)), 5&6 (TESPD(S2))) shows a strain of about 1-3% as shown in the figure (Figure 7b). In addition, as the concentration of ZB increased, it shows that the decrease in strain and increase in BO time significantly. Figure 7c shows that the addition of zinc ion to the silane/NR composite affects the increase in BO time. It also shows that the addition of ZB played a role in maximizing the above two properties.

Figure 8 shows the SEM image of the direct 3D network structure between silica-silane and NR [52]. Figure 8a shows a photograph of SEM observation that the silica surface does not have interfacial interaction with NR; however, when silane (TESPT) is added, interfacial adhesion increases between the silica surface and NR($\alpha_{FP}$) as shown in Figure 8b.

The interaction parameter in the composite material is represented by $\alpha_C$, which is expressed as $\alpha_C = \alpha_F + \alpha_P + \alpha_{FP}$. $\alpha_F$ is the filler-filler parameter in the material, $\alpha_P$ is the polymer-polymer parameter in the material, and $\alpha_{FP}$ is the filler-polymer parameter in the material [52].

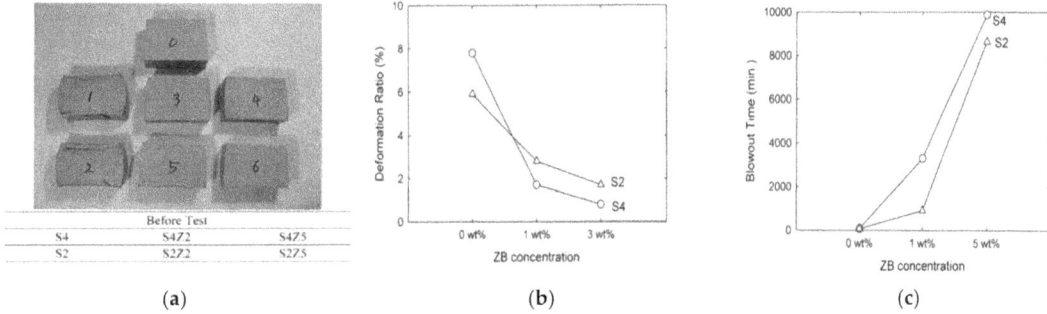

**Figure 7.** S4 (TESPT) and S2 (TESPD) compounds before and after and with/without ZB: (**a**) Photograph (0(Before Test), 1 (S4), 2 (S2), 3 (S4Z2), 4 (S4Z5), 5 (S2Z2), and 6 (SZ5)); (**b**) deformation ratio (d/D)%; and (**c**) BO time (adopted from [75]).

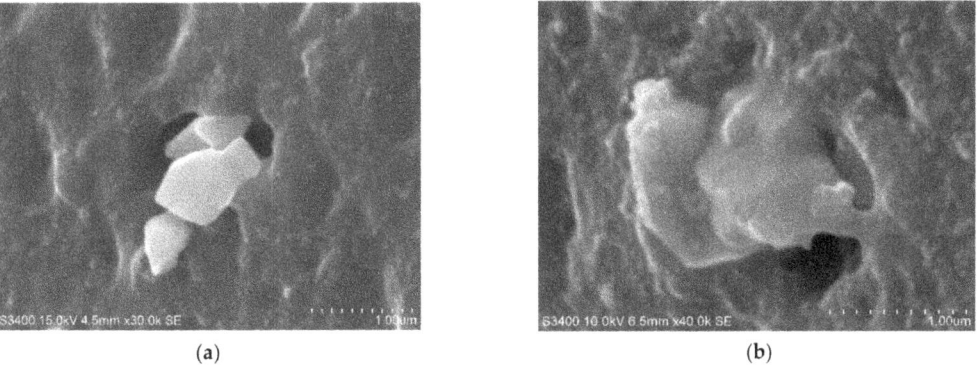

**Figure 8.** SEM photograph of silica-NR compound (**a**) without silane, (**b**) with silane(TESPT) [52].

## 5. Zinc Ion Mechanism

Zinc ions are known to affect the improvement of the 3D network structure of the silica-silane containing rubber compound, and this is explained by the reaction mechanism of iron ions [32,76,77].

Zinc is a transition element and forms a compound with the active element in the accelerator [76]. That is, due to the 'd' orbital of the zinc ion, it forms a stable divalent complex with sulfur, nitrogen, and oxygen donor elements [76]. Figure 9a shows that zinc ions form a vulcanized pre intermediate state with donor elements such as oxygen and nitrogen [32].

The interactions of accelerators and sulfur and the crosslinking are complicated mixture of reaction sequences with kinetic and thermodynamic controlling steps [74]. Both mechanisms and kinetics seem to be controlled by the zinc compound structure. At the vulcanization stage, the zinc ion increases the NR matrix network stronger. Zinc, as a transient element, is known to form complexes with accelerators [77]. Many stable divalent complexes, with sulfur, nitrogen, and oxygen donors, can form because of full 'd' orbitals on zinc [76]. The complexes are typically 4, 5, or 6 coordinated. There is also evidence that $\pi$-$\pi$ allyl complexes of zinc form between zinc and simple olefins [78]. The ligand complexes, a series of 4, 5, and 6 coordinated zinc structure explains the vulcanization process of all the coordinate sites on zinc and the electron nature of the carboxylate using expansion and contraction of the zinc orbitals [32]. The cross-linking mechanism with rubber matrix was explained by sulfur ring opening mechanism via bi-pyramidal model [32]. Two examples shown are for the sulfur ring opening reaction as shown in Figure 9a and the formation of the cross-linking as shown in Figure 9b. The sulfur ring opening reaction is shown as a bi-pyramidal complex with sulfur and oxygen anions, and sulfur and two amino ligands (residue from accelerators). The crosslinking intermediate allows for the transfer of sulfur from the polysulfide accelerator to the rubber in a combined mechanism. The π allylic model can be used to explain the carbon sulfur formation and types that form during vulcanization [74]. It also can be used to explain the differences in activation energies and reaction rates between various rubbers. This reaction binds some of the ZB onto the rubber backbone and creates a secondary ionic network.

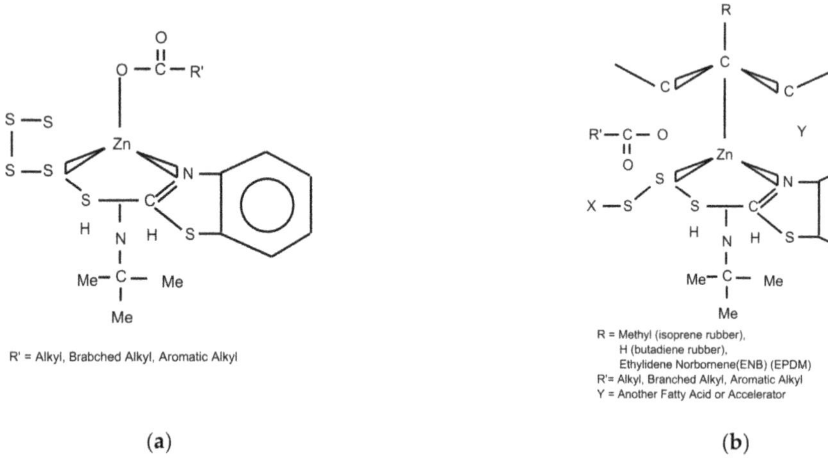

**Figure 9.** (a). Zinc Ion Effect on (a) Sulfur Ring Open Reaction Mechanism at Pre-Vulcanization Intermediate State, and (b) formation of the cross-linking intermediate (π allyl complex); adopted from [32].

Overall, "the zinc ion in zinc oxide forms a pre-intermediate state between the nitrogen (N) present in the main chain of accelerator and the oxygen (O) present on the surface of the silica to form a pre-intermediate state between the two materials. An increase in adhesion is achieved". The evidence for this is the strong interfacial adhesion between NR and silica as shown in the SEM picture in Figure 8 and the result of the increase in physical properties as shown in Figure 7b,c.

## 6. Conclusions

The vulcanization process of silica filled NR composite is as follows. The alkoxy group at one end of the silane (e.g., TESPT, TESPD) having an alkoxy function group is hydrolyzed, and the hydrolyzed silane is subjected to condensation reaction with the hydroxyl group on the silica surface. On the other side, a sulfur function group chemically bonds with the double bond of NR to form a 3D network structure between silica and rubber. At this time, under the influence of the accelerator, the degree of reaction and the rate of reaction during crosslinking are controlled. Also, the 3D network structure is greatly increased with the help of zinc ions. Figure 10 summarizes the mechanisms involved in the bonding of bifunctional alkoxy silanes with silica and NR chains and the vulcanization of zinc ions.

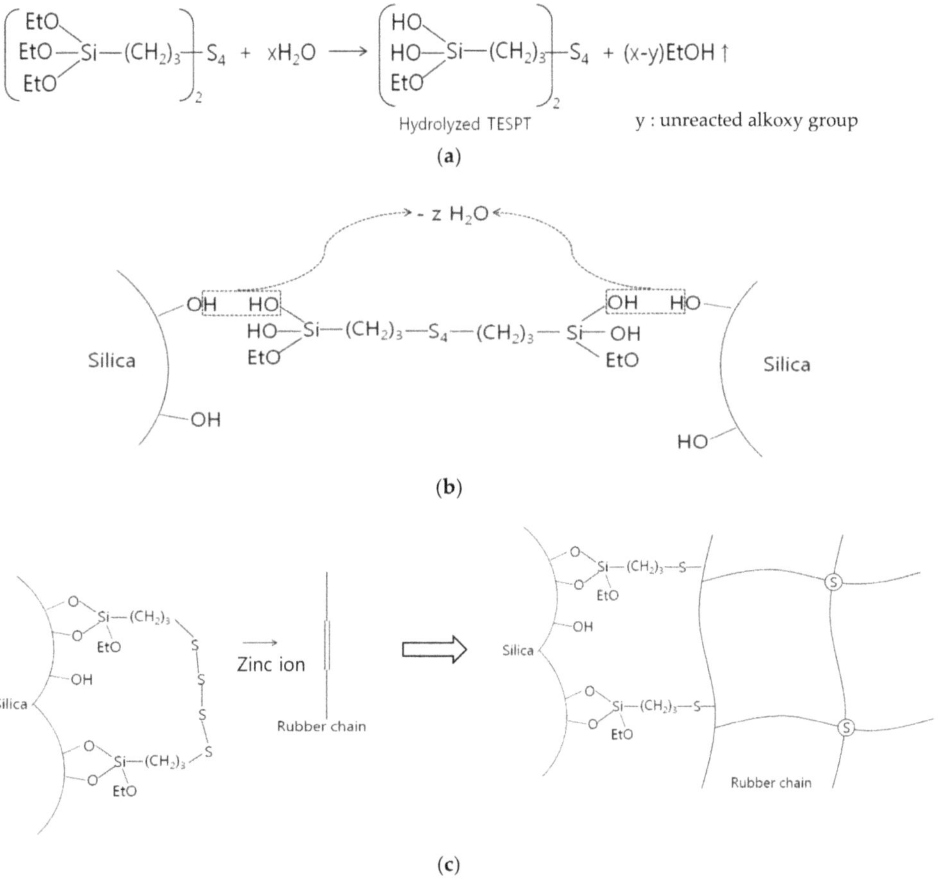

**Figure 10.** Schematic diagram of (**a**) hydrolysis of silane, (**b**) condensation between hydrolyzed silane and silica surface, (**c**) chemical reaction between sulfur and rubber chain [52].

**Author Contributions:** Conceptualization, K.-J.K.; methodology: C.S.R. All authors have read and agreed to the published version of the manuscript.

**Funding:** This research received no external funding.

**Conflicts of Interest:** The authors declare no conflict of interest.

# References

1. Kim, S.M.; Kim, K.J. Effects of accelerator on silica vs. carbon black filled natural rubber compounds. *Polym.-Korea* **2013**, *37*, 269–275. [CrossRef]
2. Goodyear, C. Improvement in India-Rubber Fabrics. U.S. Patent 3633, 15 June 1844.
3. Hancock, T. Preparation or Manufacture of Caoutchouc in Combination with Other Substances. U.K. Patent 9952, 21 November 1843.
4. Bateman, L.; Moore, C.G.; Porter, M.; Saville, B. Chapter 19. In *The Chemistry and Physics of Rubber like Substances*; Bateman, L., Ed.; John Wiley and Sons: Hoboken, NJ, USA, 1963.
5. Molony, S.B. Rubber Vulcanization and the Product Thereof. U.S. Patent 1,343,224, 9 March 1920.
6. Coran, A.Y. Chapter 7. In *Science and Technology of Rubber*, 3rd ed.; Mark, J.E., Erman, B., Eirich, F.R., Eds.; Academic Press: New York, NY, USA, 2005.
7. Bedford, C.W. Art of Vulcanizing Caoutohouc. U.S. Patent 1,371,662, 15 March 1921.
8. Sebrell, L.B.; Bedford, C.W.; Bedford, C.W. Art of Vulcanizing or Curing Caoutchouc Substances. U.S. Patent 1,544,687, 18 April 1925.
9. Bruni, G.; Romani, E. Mechanism of Action of certain Acceleratiors of Vulcanization. *Indian Rubber J.* **1921**, *62*, 63–66.

10. Harman, M.W. Process of Vulcanizing Rubber and Product Produced Thereby. U.S. Patent 2,100,692, 11 April 1933.
11. Coran, A.Y.; Kerwood, J.E. Inhibiting Premature Vulcanization of Diene Rubbers. U.S. Patent 3,546,185, 8 December 1970.
12. Thurn, F.; Burmester, K.; Pochert, J.; Wolff, S. Rubber Mixtures Giving Reversion Free Vulcanizates and Process of Vulcanization. U.S. Patent 4,517,336, 14 May 1985.
13. Rauline, R. Composition de Caoutchouc et Enveloppes de Pneumatiques à Base de Ladite Composition. European Patent EP 501,227, 2 September 1992.
14. Wagner, M.P. Reinforcing silicas and silicates. *Rubber Chem. Technol.* **1976**, *49*, 703–774. [CrossRef]
15. Gupta, R.K.; Kennal, E.; Kim, K.J. *Polymer Nanocomposites Handbook*; CRC Press: Boca Raton, FL, USA, 2009.
16. White, J.L.; Kim, K.J. *Thermoplastic and Rubber Compounds Technology and Physical Chemistry*; Hanser Publisher: Munich, Germany; Cincinnati, OH, USA, 2008.
17. Isayev, A.I.; Hong, C.K.; Kim, K.J. Continuous Mixing and Compounding of Polymer/Filler and Polymer/Polymer Mixtures with the Aid of Ultrasound. *Rubber Chem. Technol.* **2003**, *76*, 923–947. [CrossRef]
18. Kim, K.J.; White, J.L. Silica Agglomerate Breakdown In Three-Stage Mix Including A Continuous Ultrasonic Extrude. *J. Ind. Eng. Chem.* **2000**, *6*, 372–379.
19. Kim, S.M.; Cho, H.W.; Kim, J.W.; Kim, K.J. Effects of processing geometry on the mechanical properties and silica dispersion of silica-filled isobutylene-isoprene rubber (IIR) compounds. *Elastomers Compos.* **2010**, *45*, 223–229.
20. Wolff, S. Reinforcing and Vulcanization Effects of Silane Si 69 in Silica-Filled Compounds. *Kautsch. Gummi Kunstst.* **1981**, *34*, 280–284.
21. Wolff, S. Optimization of Silane-Silica OTR Compounds. Part 1: Variations of Mixing Temperature and Time during the Modification of Silica with Bis-(3-Triethoxisilylpropyl)-Tetrasulfide. *Rubber Chem. Technol.* **1982**, *55*, 967–989. [CrossRef]
22. Plueddemann, E.P. *Silane Coupling Agents*; Plenum Press: New York, NY, USA, 1982.
23. Kim, K.J.; VanderKooi, J. TESPT Treated Silica Compounds on and TESPD Rheological Property and Silica Break Down in Natural Rubber. *Kautsch. Gummi Kunstst.* **2002**, *55*, 518–528.
24. Kim, K.J.; White, J.L. TESPT and Different Aliphatic Silane Treated Silica Compounds Effects on Silica Agglomerate Dispersion and on Processability During Mixing in EPDM. *J. Ind. Eng. Chem.* **2001**, *7*, 50–57.
25. Kim, K.J. Bifunctional silane (TESPD) effects on silica containing elastomer compound: Part I: Natural Rubber (NR). *Elastomers Compos.* **2009**, *44*, 134–142.
26. Jeon, D.K.; Kim, K.J. Bifunctional silane (TESPD) effects on silica containing elastomer compound: Part II: Styrene-co-Butadiene Rubber (SBR). *Elastomers Compos.* **2009**, *44*, 252–259.
27. Kim, K.J. Bifunctional silane (TESPD) effects on improved mechanical properties of silica containing nitrile-butadiene rubber/poly(vinyl chloride) compound. *J. Appl. Polym. Sci.* **2012**, *124*, 2937. [CrossRef]
28. Dang, T.T.N.; Kim, J.K.; Kim, K.J. Concentration effects of organosilane (TESPD) on mechanical properties of silica filled Silicone Rubber/Natural Rubber compounds. *Int. Polym. Proc.* **2011**, *26*, 368–374. [CrossRef]
29. Kim, K.J. Amino silane, vinyl silane, TESPD, ZS (TESPD/zinc complex) effects on carbon black/clay filled chlorobutyl rubber (CIIR) compounds; Part I: Effects on hard clay/carbon black filled compounds. *Carbon Lett.* **2009**, *10*, 101–108. [CrossRef]
30. Kim, K.J. Amino silane, vinyl silane, TESPD, ZS (TESPD/zinc complex) effects on carbon black/clay filled chlorobutyl rubber (CIIR) compounds; Part II: Effects on soft clay/carbon black filled compounds. *Carbon Lett.* **2009**, *10*, 109–113. [CrossRef]
31. Kim, K.J. Amino silane, vinyl silane, TESPD, ZS (TESPD/zinc complex) effects on carbon black/clay filled chlorobutyl rubber (CIIR) compounds; Part III: Comparative studies on hard clay and soft clay filled compounds. *Carbon Lett.* **2009**, *10*, 190–197. [CrossRef]
32. Kim, K.J.; VanderKooi, J. Rheological Effects of Zinc Surfactant on the TESPT-Silica Mixture in NR and S-SBR Compounds. *Int. Polym. Proc.* **2002**, *17*, 192–200. [CrossRef]
33. Kim, K.J.; VanderKooi, J. Zinc Surfactant Effects on Processability and Mechanical Properties of Silica Filled Natural Rubber Compounds. *J. Ind. Eng. Chem.* **2004**, *10*, 772–781.
34. Kim, K.J. Ethylene–propylene–diene terpolymer/silica compound modification with organosilane [bis(triethoxysilylpropyl)disulfide] and improved processability and mechanical properties. *J. Appl. Polym. Sci.* **2010**, *116*, 237–244. [CrossRef]
35. Dang, T.T.N.; Kim, J.K.; Kim, K.J. Zinc Surfactant Effects on Improved Mechanical Properties of SR/NR/Silica/TESPD Compounds. *Int. Polym. Proc.* **2009**, *24*, 359–367. [CrossRef]
36. Kim, K.J.; VanderKooi, J. Moisture Effects on Improved Hydrolysis Reaction for TESPT and TESPD-Silica Compounds. *Compos. Interfaces* **2004**, *11*, 471–488. [CrossRef]
37. Kim, K.J.; VanderKooi, J. Moisture Effects on TESPDSilica/CB/SBR Compounds. *Rubber Chem. Technol.* **2005**, *78*, 84–104. [CrossRef]
38. Kim, K.J.; VanderKooi, J. Temperature Effects of Silane Coupling on Moisture Treated Silica Surface. *J. Appl. Polym. Sci.* **2005**, *95*, 623–633. [CrossRef]
39. Kim, S.M.; Nam, C.S.; Kim, K.J. TMTD, MBTS, and CBS accelerator effects on a silica filled natural rubber compound upon vulcanization properties. *Appl. Chem. Eng.* **2011**, *22*, 144–148.
40. Choi, C.Y.; Kim, S.M.; Park, Y.H.; Jang, M.K.; Nah, J.W.; Kim, K.J. Effects of Thiuram, thiazole, and sulfenamide accelerators on silica filled natural rubber compound upon vulcanization and mechanical properties. *Appl. Chem. Eng.* **2011**, *22*, 411–415.

41. Dang, T.T.N.; Kim, J.K.; Lee, S.H.; Kim, K.J. Vinyl functional group effects on mechanical and thermal properties of silica-filled Silicone Rubber/Natural Rubber blends. *Compos. Interf.* **2011**, *18*, 151–168. [CrossRef]
42. Dang, T.T.N.; Kim, J.K.; Kim, K.J. Organo bifunctional silane effects on the vibration, thermal, and mechanical properties of a vinyl-group-containing silicone rubber/natural rubber/silica compound. *J. Vinyl Addit. Technol.* **2010**, *16*, 254–260. [CrossRef]
43. Lorenz, O.; Echte, E. The Vulcanization of Elastomers. 13. The Vulcanization of Natural Rubber with Sulfur in the Presence of Mercaptobenzothiazole. II. *Rubber Chem. Technol.* **1958**, *31*, 117–131. [CrossRef]
44. Scheele, W.; Cherubim, M. Vulcanization of Elastomers. 30. Kinetics of the Decrease of Sulfur Concentration during Vulcanization. *Rubber Chem. Technol.* **1961**, *34*, 606–628. [CrossRef]
45. Morita, E.; Young, E.J. A Study of Sulfenamide Acceleration. *Rubber Chem. Technol.* **1963**, *36*, 844–862. [CrossRef]
46. Bhatnagar, S.K.; Banerjee, S. Kinetics of Accelerated Vulcanization—III N-Cyclohexyl-Benzothiazole-2-Sulfenamide Accelerated Sulfur Vulcanization of Rubbers. *Rubber Chem. Technol.* **1969**, *42*, 1366–1382. [CrossRef]
47. Wolff, S. Empirisch ermittelte Zusammenhange Zwischen Kautschuk/Fullstoff-Wechselwirkung und Prufwerten statischer und dynamischer Vulkanisat-Untersuchungen. *Kautsch. Gummi. Kunstst.* **1969**, *22*, 367.
48. Wolff, S. Moeglichkeit einer neuen Charakterisierrung der Traction Resistance, Brevet Wirkungsweise von Russen in 1, 5-Polyenen. *Kautsch. Gummi. Kunstst.* **1970**, *23*, 7–14.
49. Wolff, S. Chemical Aspects of Rubber Reinforcement by Fillers. *Rubber Chem. Technol.* **1996**, *69*, 325–346. [CrossRef]
50. Wolff, S.; Wang, M.J. Filler—Elastomer Interactions. Part IV. The Effect of the Surface Energies of Fillers on Elastomer Reinforcement. *Rubber Chem. Technol.* **1992**, *65*, 329–342. [CrossRef]
51. Tan, E.H.; Wolff, S.; Haddeman, M.; Grewatta, H.P.; Wang, M.J. Filler—Elastomer Interactions. Part IX. Performance of Silicas in Polar Elastomers. *Rubber Chem. Technol.* **1993**, *66*, 594–604. [CrossRef]
52. Kim, K.J. Silane effects on in-rubber silica dispersion and silica structure (alpha(F)): A Review. *Asian J. Chem.* **2013**, *25*, 5119–5123. [CrossRef]
53. Kim, K.J. Overview of Hydrolysis: A Review Part I- Hydrolysis Mechanism. *Elast. Compos.* **2020**, *55*, 128–136.
54. Kim, K.J. Overview of Hydrolysis: A Review Part II- Hydrolysis Application. *Elast. Compos.* **2020**, *55*, 137–146.
55. Arkles, B.; Steinmetz, J.R.; Zazyczny, J.; Metha, P. *Silanes and Other Coupling Agents*; Mittal, K.L., Ed.; VSP: Utrecht, The Netherlands, 1992; p. 93.
56. Oral, U.; Hunsche, A. Advanced investigations into the silica/silane reaction system. *Soc. Rubber Ind.* **1998**, *71*, 549–561.
57. Oral, U.; Hunsche, A.; Müller, A.; Koban, H.G. Investigations into the Silica/Silane Reaction System. *Rubber Chem. Technol.* **1997**, *70*, 608–623.
58. Hunsche, A.; Gorl, U.; Müller, A.; Knaack, M.; Göbel, T. Investigations concerning the reaction silica/organosilane and organosilane/polymer. Part 1: Reaction mechanism and reaction model for silica/organosilane. *Kautsch. Gummi Kunstat.* **1997**, *50*, 881–889.
59. Lee, J.Y.; Kim, K.J. MEG effects on hydrolysis of polyamide 66/glass fiber composites and mechanical property changes. *Molecules* **2019**, *24*, 755. [CrossRef] [PubMed]
60. Taft, R.W., Jr. Chapter. 13. In *Steric Effects in Organic Chemistry*; Newman, M.S., Ed.; Wiley: New York, NY, USA, 1956.
61. McNeil, K.J.; DiCapri, J.A.; Walsh, D.A.; Pratt, R.F. Kinetics and mechanism of hydrolysis of a silicate triester, tris(2-methoxyethoxy)phenylsilane. *J. Am. Chem. Soc.* **1980**, *102*, 1859–1865. [CrossRef]
62. Vorokonov, M.G.; Meleshkevisk, V.P.; Yuzekelvski, Y.A. *The Siloxane Bond*; Plenum Press: New York, NY, USA, 1978; pp. 375–380.
63. DeTar, D.F. Effects of alkyl groups on rates of acyl-transfer reactions. *J. Org. Chem.* **1980**, *45*, 5166–5174. [CrossRef]
64. Jencks, W.P.; Salvesen, K. Equilibrium deuterium isotope effects on the ionization of thiol acids. *J. Am. Chem. Soc.* **1971**, *93*, 4433–4436. [CrossRef]
65. Prassas, M.; Hench, L.L. *Ultrastructure Processing of Ceramics*; Hench, L., Ulrich, D., Eds.; John Wiley: New York, NY, USA, 1984; p. 100.
66. Allen, K.W. *Silane Coupling Agents*; Mittal, K.L., Ed.; VSP: Utrechi, The Netherlands, 1992.
67. Kay, B.D.; Assink, R.A. Sol-gel kinetics: II. Chemical speciation modeling. *J. Non-Cryst. Solids* **1988**, *104*, 112–122. [CrossRef]
68. Osterholtz, F.D.; Pohl, E.P. *Silane Coupling Agents*; Mittal, K.L., Ed.; VSP: Utrechi, The Netherlands, 1992.
69. Moore, M.J. Silanes improve rubber-to-metal bonding. *Rubber Plast. News* **2002**, *31*, 14–16.
70. Kim, K.J.; VanderKooi, J. Reactive Batch Mixing for Improved Silica-Silane Coupling. *Int. Polym. Process.* **2004**, *19*, 364–373. [CrossRef]
71. Hashim, A.S.; Zahare, B.A.; Ikeda, Y.; Ohjiya, S.K. The Effect of Bis(3-Triethoxysilylpropyl) Tetrasulfide on Silica Reinforcement of Styrene-Butadiene Rubber. *Rubber Chem. Technol.* **1998**, *78*, 289–299. [CrossRef]
72. Heiken, D.; Barentsen, W. Particle dimensions in polystyrene/polyethylene blends as a function of their melt viscosity and of the concentration of added graft copolymer. *Polymer* **1977**, *18*, 69–72. [CrossRef]
73. Lee, J.Y.; Kim, K.J. Overview of Polyamide Resins and Composites: A Review. *Elast. Compos.* **2016**, *51*, 317–341. [CrossRef]
74. Roberts, A.D. Chapter 12. In *Natural Rubber Science and Technology*; Oxford Science Publishers: New York, NY, USA, 1988.
75. Kim, K.J.; VanderKooi, J. Effects of Zinc Ion Containing Surfactant on Bifunctional Silane Treated Silica Compounds in Natural Rubber. *J. Ind. Eng. Chem.* **2002**, *8*, 334–347.
76. Farnsworth, M.; Kline, C. *Zinc Chemicals*; Chales Kline and Co.: New York, NY, USA, 1983.

77. Duchacek, V.; Kuta, A.; Pribyl, P. Efficiency of metal activators of accelerated sulfur vulcanization. *J. Appl. Polym. Sci.* **1993**, *47*, 743–746. [CrossRef]
78. Kokes, R. Catalysis of Hydrocarbons and Related Studies. *Intra-Sci. Rept.* **1972**, *6*, 77.

Article

# Simultaneous Effects of Carboxyl Group-Containing Hyperbranched Polymers on Glass Fiber-Reinforced Polyamide 6/Hollow Glass Microsphere Syntactic Foams

Jincheol Kim [1,†], Jaewon Lee [1,†], Sosan Hwang [1], Kyungjun Park [1], Sanghyun Hong [2], Seojin Lee [2], Sang Eun Shim [1,\*] and Yingjie Qian [1,\*]

[1] Department of Chemistry and Chemical Engineering, Education and Research Center for Smart Energy and Materials, Inha University, Michuhol-gu, Incheon 22212, Korea; 22192178@inha.edu (J.K.); 22192186@inha.edu (J.L.); bulls@inha.edu (S.H.); 22201542@inha.edu (K.P.)

[2] LG Electronics Inc. 51, Gasan Digital 1-ro, Geumcheon-gu, Seoul 08592, Korea; sanghyun.hong@lge.com (S.H.); seojin710.lee@lge.com (S.L.)

\* Correspondence: seshim@inha.ac.kr (S.E.S.); yjqian@inha.ac.kr (Y.Q.)

† These authors contributed equally to this work.

**Abstract:** The hollow glass microsphere (HGM) containing polymer materials, which are named as syntactic foams, have been applied as lightweight materials in various fields. In this study, carboxyl group-containing hyperbranched polymer (HBP) was added to a glass fiber (GF)-reinforced syntactic foam (RSF) composite for the simultaneous enhancement of mechanical and rheological properties. HBP was mixed in various concentrations (0.5–2.0 phr) with RSF, which contains 23 wt% of HGM and 5 wt% of GF, and the rheological, thermal, and mechanical properties were characterized systematically. As a result of the lubricating effect of the HBP molecule, which comes from its dendritic architecture, the viscosity, storage modulus, loss modulus, and the shear stress of the composite decreased as the HBP content increased. At the same time, because of the hydrogen bonding among the polymer, filler, and HBP, the compatibility between filler and the polymer matrix was enhanced. As a result, by adding a small amount (0.5–2.0 phr) of HBP to the RSF composite, the tensile strength and flexural modulus were increased by 24.3 and 9.7%, respectively, and the specific gravity of the composite was decreased from 0.948 to 0.917. With these simultaneous effects on the polymer composite, HBP could be potentially utilized further in the field of lightweight materials.

**Keywords:** syntactic foams; hyperbranched polymer; polyamide 6; hollow glass microsphere; lubricant; compatibilizer; composites

## 1. Introduction

The international automobile market has been changing from traditional internal combustion engines to eco-friendly vehicle development, in line with corporate average fuel efficiency (CAFÉ) standards and automobile greenhouse gas (GHG) emission standards [1]. As economic and the environmental concerns for fuel consumption have evolved, lightweight materials became of great interest to the automotive industry [2–7]. In order to apply lightweight materials in the automotive industry, the mechanical strength and the processability of the material must be accompanied by reduced specific gravity.

To reduce the specific gravity of the polymeric composite materials, numerous studies have been actively conducted on hollow glass microsphere (HGM)-containing lightweight syntactic foams (SFs) in various applications, such as automotive, marine, and aerospace [8–15]. HGM exhibits a low specific gravity and high electrical and thermal insulation properties. However, the addition of HGMs in higher contents results in the weakening of mechanical properties of SFs [12,16]. Addition of fibrous material, such as glass fiber (GF) or carbon fiber to SFs, can enhance the mechanical properties of the SFs;

these are named reinforced syntactic foam (RSF) [17–19]. Nevertheless, the addition of rigid HGM and GF particles have a detrimental effect on the processability by increasing viscosity. Furthermore, due to the high shear of the extrusion process, HGM particles are broken, and this breakage results in an increase in the specific gravity of the polymer composites [9,20].

To address these issues, lubricating agents and plasticizers have been widely studied [21–24]. These materials modify the viscosity of the polymer melt, and reduce the friction; as a result, they increase the processability of the polymer composite with their addition in small amounts to the composites. The paraffin waxes, esters, and fatty acids derivatives are commonly used as lubricants for polymers [25]. Generally, when these typical lubricants are added to the composite material, the mechanical strength of the material weakens, since these lubricants have low molecular weights and low compatibility with polymer matrices [26,27]. Recently, topology-engineered polymers, such as star-shaped polymers and hyperbranched polymers have been studied, in order to be applied as lubricants [24,28].

Hyperbranched polymer (HBP) is a highly branched three-dimensional polymer. The HBP molecule has a low degree of entanglement and low viscosity due to the dendritic architecture of the molecule [29,30]. It can be utilized as a rheological modifier in polymeric composites by acting as molecular ball-bearings at the expense of the van der Waals forces [31]. According to WanG's study [24], as the polymer chain topology changes from a linear to a hyperbranched structure, the intrinsic viscosity and the complex viscosity of the polymer decrease significantly, and the shear stability of the polymer increases as well. Besides, the abundant functional groups in the HBP molecules make it possible to form hydrogen bonding among HBP, polymer matrix, and the fillers in the composite material [32,33]. This functionality results in the compatibilization of the filler in the composite material, and could make up for the weakened mechanical strength of the syntactic foams. Gu et al. [34] increased the toughness and the mechanical strength of soybean protein film with addition of hyperbranched polyester, which forms strong hydrogen bonding with the soybean protein matrix. Peng et al. [35] improved the wettability and the interfacial adhesion of carbon fiber to an epoxy resin matrix by poly(amido amine) functionalization.

As a result of its dendritic structure and abundant functional groups, HBP could act as a lubricating additive and as a compatibilizer at the same time in the GF-reinforced SF system. In the present study, the carboxyl group-containing HBP molecule was introduced to RSF composite material to reduce the viscosity and to improve the mechanical properties of the composite material simultaneously. We used polyamide 6 (PA 6) as a polymer matrix, since PA 6 is widely used in the automotive industry, and HBP can be dispersed effectively in the PA 6 matrix due to the hydrogen bonding ability of PA 6. Scanning electron microscopy (SEM) microphotographs revealed that the compatibility between filler and PA 6 polymer matrix was enhanced with the addition of a small amount of HBP (0.5–2.0 phr). The density of the RSF and ashes of the RSF were measured to calculate HGM breakage of RSF material. Tensile strength increased significantly from 59 to 73 MPa, and the specific gravity decreased from 0.948 to 0.917 with the addition of HBP.

## 2. Materials and Methods

*2.1. Materials*

Injection grade Nylon-6 (Ultramid, B3S, Ludwigshafen, Germany), with a density of 1.13 g/cm$^3$ and melt volume rate of 160 cm$^3$/min at a temperature of 275 °C, given load of 5 kg, was supplied by BASF (Ludwigshafen, Germany). HGM (IM16K) provided by 3M (Saint Paul, SP, USA), with a true density of 0.46 g/cm$^3$, a crush strength of 110 MPa, and average diameter of 20 μm, was used as the lightweight filler in this study. The chopped nylon-compatible-sized GF used for a reinforced filler was 995-10P grade from Owens Corning (Toledo, OH, USA). The GF grade has a nominal diameter of 10 μm, chopped length of 4 mm, and sized with amino silane coupling agent. Carboxyl group functionalized

HBP, CYD-7010 (synthesized from adipic acid and hyperbranched polyester with 98:2 molar ratio), with a melt range of 135~155 °C, was supplied by Weihai CY Dendrimer Technology Corporation (Weihai, China). The chemical structures of PA 6, HGM, GF, and HBP are presented in Figure 1.

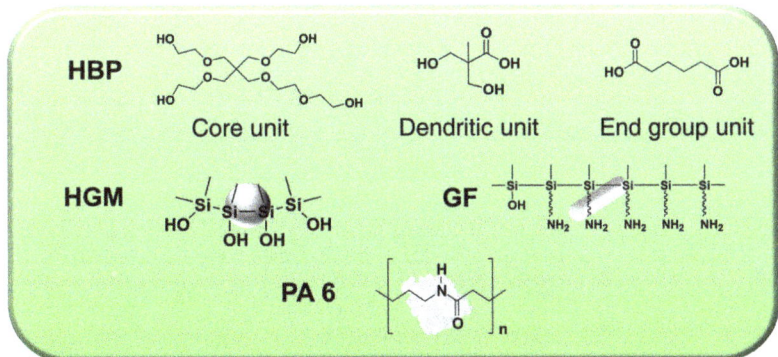

**Figure 1.** Chemical structures of PA 6, HGM, GF, and HBP.

## 2.2. Methods

### 2.2.1. Sample Preparation

The syntactic foams were prepared in a co-rotating twin-screw extruder (TEK-25, SM PLATEC Co., Ansan, Korea), with screw diameter of 25 mm, $D_o/D_i = 1.55$, L/D = 40, comprising two kneading zones for both dispersive and distributive mixing. Twin screw configuration is depicted in Figure S1. In order to minimize hydrolysis during the extrusion process, Nylon-6 was pre-dried at 80 °C for 24 h. The PA 6 pellets and HBP powder were fed into the main hopper first. Then, the HGM was fed into the first side feeder, after which the GF was fed into the second side feeder. All composites were melt-mixed at the same conditions under 200 rpm at a temperature range of 240/240/250/250/250/250/250/250/250/250 °C from hopper to die. The PA 6 and the HBP were pre-melted before introducing HGM fillers. The extrudates were passed through a hot water bath, pelletized, and dried at 80 °C for 24 h before injection molding. Specimens for the tensile, flexural testing, and density measurements were injection molded in a lab-scale injection molding machine (VDCII-50, JINHWA GLOTECH, Cheonan, Korea) with a clamping force of 50 tons. Injection molding was carried out at a barrel temperature profile of 190/220/230/240/240 °C from hopper to nozzle, and a mold temperature of 80 °C. The compositions and the codes of the prepared samples are reported in Table 1.

**Table 1.** Composition and identification details of prepared samples referred to as syntactic foams (SF), consisting of different PA 6, HGM, GF, and HBP contents.

| Sample Label | PA 6 (wt%) | HGM (wt%) | GF (wt%) | HBP (phr [a]) |
|---|---|---|---|---|
| RSF | 72 | 23 | 5 | - |
| RSF-COOH 0.5 | 71.6 | 23 | 5 | 0.5 |
| RSF-COOH 0.75 | 71.5 | 23 | 5 | 0.75 |
| RSF-COOH 1.0 | 71 | 23 | 5 | 1.0 |
| RSF-COOH 2.0 | 70.6 | 23 | 5 | 2.0 |

[a] Parts per hundred of PA 6 resin.

### 2.2.2. Characterization

Complex viscosity and shear modulus were measured using an Anton Paar MCR 302 rheometer (Graz, Austria) with 25-mm parallel plate geometry. Extrudates of neat Nylon-6 and composites were molded at 240 °C. Small angle oscillatory shear (SAOS) frequency sweep tests were performed at a constant temperature of 240 °C within a linear viscoelastic regime in the range of 0.1–100 rad/s.

The tensile-fractured samples for morphological observation were characterized with field-emission scanning electron microscopy (FE-SEM) using the model S-4300 from HI-TACHI (Tokyo, Japan). The fractured surface was sputter-coated with platinum. Right after the tensile testing, the fractured surface of each sample was dipped in liquid nitrogen for 10 min to keep the shape of the surface.

The differential scanning calorimetry (DSC) analysis was conducted with NETZSCH DSC200F3 (Selb, Germany), with a sample mass of 3 mg that was set in an aluminum pan with a cover under an $N_2$ atmosphere with a flow rate of 40 mL/min, to determine the crystallization behavior of the composites. The glass transition temperature ($T_g$), melting temperature ($T_m$), and melt crystallization temperature ($T_c$) were measured from the second heating and cooling cycle, respectively. The profile of the thermogram is depicted in Figure S2. The samples were first heated from 20 to 250 °C at a rate of 30 °C/min and held at 250 °C for 4 min to eliminate thermal history of the samples. The samples were then cooled down to −10 °C at a rate of 30 °C/min and held at −10 °C for 10 min. The DSC data were collected in the second cycle (segments 6 and 8 as heating and cooling, respectively). The samples were heated at a rate of 10 °C/min from −10 to 250 °C, and then held at 250 °C for 4 min. Then the samples were cooled to −10 °C at a rate of 10 °C/min. The samples were finally held at −10 °C for 4 min.

The crystallinity of the samples was calculated by the following equation:

$$X_c\ (\%) = \frac{\Delta H_f}{W_m \cdot \Delta H_o} \quad (1)$$

where $\Delta H_f$ is heat of fusion for the sample, $W_m$ is the mass fraction of the PA 6, and the $\Delta H_o$ is the heat of fusion of the theoretical 100% crystalline PA 6 (240 J/g).

The non-isothermal crystallization behavior of the composites was calculated using the Avrami equation [36–38]:

$$1 - V_t = \exp(-Z_t t^n) \quad (2)$$

where $V_t$ is the relative volumetric fraction of crystalline, $Z_t$ is the overall crystallization rate constant which reflects both nucleation and crystal growth, and $n$ is the Avrami index. The $V_t$ can be obtained by the following equation:

$$V_t = \frac{W_c}{W_c + \left(\frac{\rho_c}{\rho_a}\right)(1 - W_c)} \quad (3)$$

where $W_c = \Delta H(t)/\Delta H_{total}$ is the crystalline mass fraction, and $\rho_c$ and $\rho_a$ are crystalline density (1.20 g/cm$^3$) and amorphous density (1.09 g/cm$^3$) of the PA 6, respectively [39]. When the crystallization rate constant is corrected by taking the cooling rate into account for the Avrami equation, which was applied only to the isothermal crystallization, the following equation is used, where the $Z_c$ is the crystallization rate constant in the non-isothermal condition:

$$\log Z_c = \frac{\log Z_t}{dT/dt} \quad (4)$$

Thermogravimetric analysis (TGA) was performed with PerkinElmer TGA4000 (Waltham, MA, USA) under an $N_2$ atmosphere with a flow rate of 20 mL/min. The samples with a mass of 10 mg were contained in an aluminum pan, and measured at a heating rate of 10 °C/min from 40 to 800 °C to determine thermal stability of the composite samples.

The tensile strength (ASTM D638) and flexural modulus (ASTM D790) of each specimen were performed using a universal testing machine (DUT-2TC, DAEKYUNG ENGINEERING Co., Bucheon, Korea) with a 2-ton load cell. Specimen dimensions for the tensile test are presented in Figure S1b. Samples for the flexural test having dimensions of 127 × 12.7 × 6.4 mm$^3$ (length × width × thickness) were employed for the test in accordance with ASTM D790. The tensile tests were performed at a cross-head speed of 50 mm/min. The flexural tests were carried out in the three-point bending configuration at a cross-head speed of 1.54 mm/min. Flexural modulus was calculated from the slope of the initial linear portion of the curves. The five different samples were tested for accurate results. Two types of density measurements were conducted with a hydro-densimeter (GP300S, MATSUHAKU, Taichung, Taiwan) and a gas pycnometer (BELPycno, Micro-tracBEL Corp., Osaka, Japan) to measure the bulk density of the composites and residue inorganic ashes of the composites, respectively.

## 3. Results and Discussion

### 3.1. Rheological Properties

The rapid increase in complex viscosity in RSF control is shown in Figure 2a due to the high content of fillers (HGM, GF). As the contents of HBP increased, the complex viscosity of the composites was reduced across the entire frequency range. It was confirmed that the complex viscosity of RSF-COOH 2.0 decreased by 4.9 times at 0.1 rad/s compared to the neat RSF due to the addition of the HBP molecules in the composite system. Figure 2b shows that the shear stress decreased as the HBP content increased. The internal friction was alleviated by HBP, which has a low intrinsic viscosity in the molten state. HBP molecules improved the polymer chain mobility due to its lubricating effect, and this improved polymer chain mobility in filled RSF composite is on account of the dendritic architecture of HBP [40]. The reduced shear stress could contribute to the reduction in HGM breakage arising from the high shear in the extrusion process and decrease in the specific gravity of the composite.

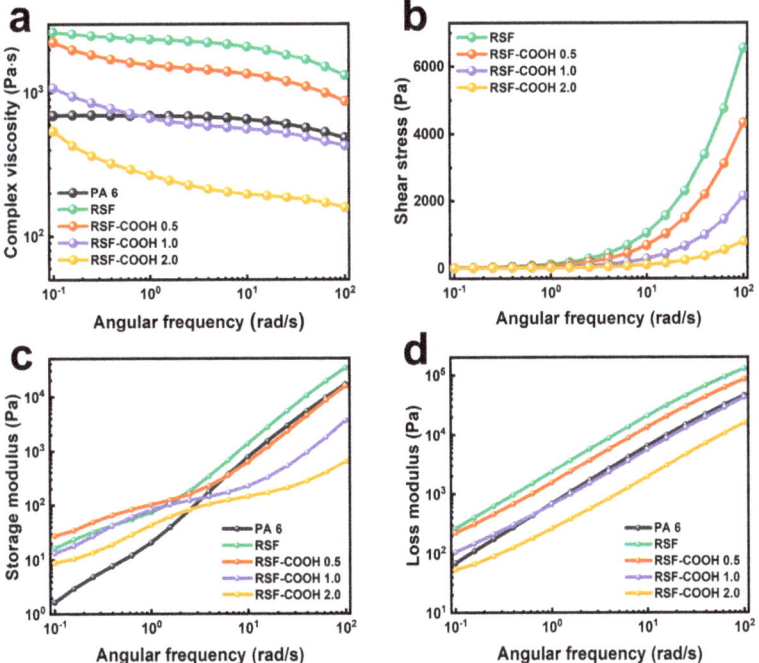

**Figure 2.** Complex viscosity (**a**), shear stress (**b**), storage modulus (**c**), and loss modulus (**d**) of the neat PA6 and RSF composites with different HBP contents.

Changes in storage modulus (G′) and loss modulus (G″) of the composites for different contents of HBP in RSF are shown in Figure 2c,d. In overall frequency, the G″ was higher than the G′ value, which means the composites were measured in the predominantly viscous region. Kang et al. [41] found that the syntactic foams aggregated at higher content (higher than HGM 10 wt%), and the viscoelastic response changed from viscous to elastic behavior at low frequency. Similarly, both the storage modulus (G′) and the loss modulus (G″) of the RSF were increased compared to the neat PA 6. When HBP was added to the RSF, the G′ and G″ decreased in overall frequency. Due to the low entanglement degree of the HBP molecule itself, the lubricating effect was applied in the composite system; as a result, the overall G′ and G″ tend to decrease in RSF-COOH composites. At high frequencies, the alteration of segmental dynamics was not observed because the entanglement structure of the polymer was retained [42]. As a result, the effect of the lubrication which comes from the dendritic architecture of the HBP molecule was predominant. However, at low frequencies the slope of the G′ to the frequency was decreased in all samples compared to the RSF, and the G′ value slightly increased when 0.5 phr of HBP was added. These phenomena come from the abundant functional groups in the HBP molecule. When the entanglement of the polymer chain was released at low frequency, the unravelling of the entanglement was hindered by the hydrogen bonding among the PA 6, fillers, and the HBP molecules. A similar result was observed in Bhardwaj's research [43]. This could be evidence that not only hydrogen bonding, but also lubrication, were functioning in the RSF-COOH composite system.

## 3.2. Morphology

SEM microphotographs of the fractured surfaces of the composite materials after tensile testing are presented in Figure 3. In Figure 3a, without HBP, most of the HGMs are exposed to the fractured surface. However, when the HBP was added to the RSF (Figure 3b–d), the partially exposed and almost buried HGMs were observed, indicating that HBP increased the interfacial adhesion between polymer matrix and HGM. The enlarged microphotographs of the GFs are presented in Figure 3e–h. The GF surface of the neat RSF was smooth without attached PA 6 matrix. On the other hand, when HBP was added, the GF surface was covered with PA 6. On account of the enhanced interfacial adhesion of the polymer matrix to the fillers, which comes from the hydrogen bonding among the polymer matrix, fillers, and HBP, the compatibility of the fillers in the PA 6 matrix was increased, and the mechanical strength of the RSF-COOH composites could be enhanced compared to the neat RSF.

It would be useful to discuss the fracture mechanism of the composite material to investigate the change of the mechanical properties of the composites with and without HBP addition. When the external load was applied to the composite, micro-cracks were generated near the filler surface due to weak interfacial adhesion with the polymer matrix [18]. Without HBP, when a high content of HGM was added, the sites of cracks increased; furthermore, the cracks propagated easily because there were neighboring cracks from other HGM particles. Since the strength of the matrix is larger than that of the interface between the HGM and polymer matrix [44], cracks on the surface of the HGM propagated easily. On the other hand, when HBP was added to the RSF composite, the filler-polymer adhesion increased due to the increased hydrogen bonding sites that came from the abundant functional groups of the HBP molecule. As a result, under the external load, the effect of the cracks on the surface of the HGMs decreased and the effect of the plastic deformation and yielding became predominant. Therefore, the mechanical strength of the RSF-COOH was higher than that of the neat RSF composite.

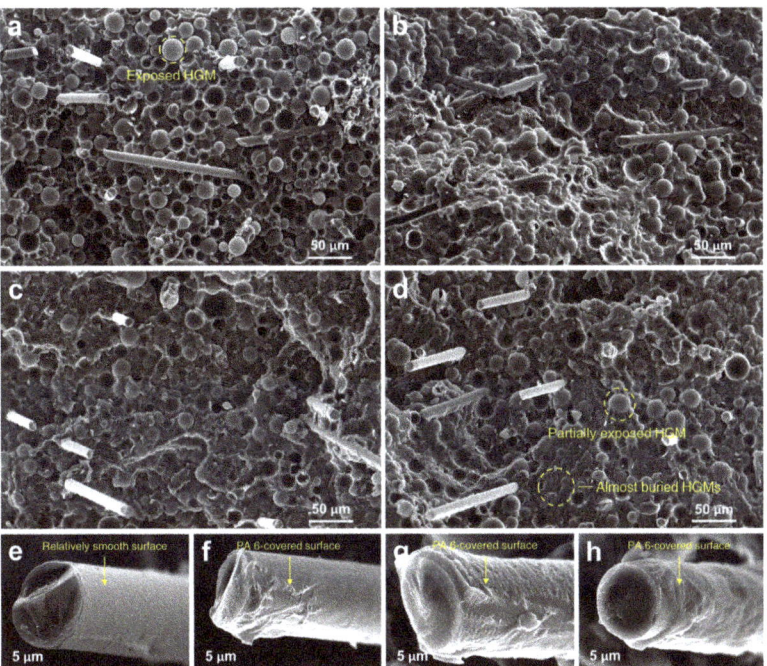

**Figure 3.** SEM microphotographs of fractured surface of the composite samples after tensile testing: (**a**) RSF control, (**b**) RSF-COOH 0.5, (**c**) RSF-COOH 1.0, and (**d**) RSF-COOH 2.0. Enlarged microphotographs of the GFs which are in the (**e**) RSF control, (**f**) RSF-COOH 0.5, (**g**) RSF-COOH 1.0, and (**h**) RSF-COOH 2.0, respectively.

*3.3. Crystallization Behavior*

The cooling curves and the DSC data of the neat PA 6, PA 6/HGM, and PA 6/GF composites at temperatures ranging from 205 to 180 °C are shown in Figure S5 and Table S1. The crystallization temperature of the PA 6/HGM 20 composite decreased compared to the neat PA 6. However, the change in crystallization temperature of the PA 6/HGM 5 and the PA 6/GF 5 samples was not significant. However, high content of HGM particles, which was associated with our RSF system, influenced the crystallization behavior significantly. The crystallization rate constant $Z_t$ decreased, and the $t_{1/2}$ of the PA 6/HGM 20 increased. These phenomena reflect that the crystallization of the RSF without HBP was delayed since the HGM particles in the PA 6 matrix could act as the steric obstacle for the crystalline development of the PA 6 polymer [45]. The $T_g$ data of the PA 6, RSF, and RSF-COOH samples are presented in Figure S6. The $T_g$ of the samples decreased from 60.81 to 45.52 °C as the HGM and GF were introduced to the PA 6; this was due to the weak interfacial adhesion between the fillers and PA 6 [46]. However, when HBP was added to the RSF system, the $T_g$ of the composites slightly increased from 45.52 to 46.70 °C. This indicates that HBP addition enhanced the interfacial adhesion of the fillers with the PA 6 matrix.

Figure 4a presents the DSC thermograms of the neat RSF and RSF-COOH composites, and the DSC data are listed in Table 2.

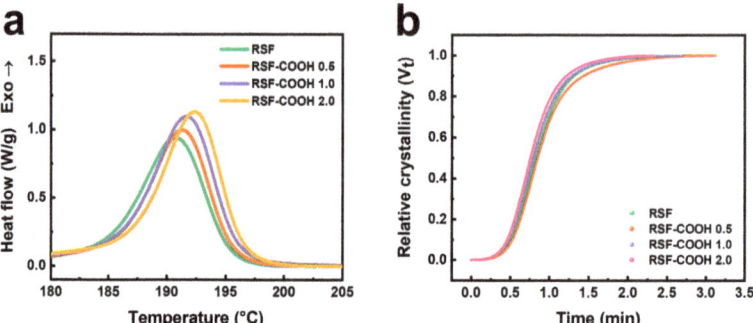

**Figure 4.** DSC thermograms cooled at 10 °C/min (**a**) and plots of relative crystallinity ($V_t$) vs time (**b**) for non-isothermal crystallization of RSF composites with various HBP contents.

**Table 2.** DSC data for RSF containing various HBP contents under non-isothermal crystallization.

| Sample Label | $T_m$ (°C) | $\Delta H_f$ (J/g) | $X_c$ (%) | $T_c$ (°C) | $T_{onset}$ (°C) | $\Delta H_c$ (J/g) | $\Delta H_c^*$ (J/g) |
|---|---|---|---|---|---|---|---|
| RSF | 221.3 | 34.68 | 20.00 | 190.8 | 195.1 | 40.98 | 56.70 |
| RSF-COOH 0.5 | 220.8 | 35.00 | 20.40 | 191.3 | 195.4 | 41.16 | 57.57 |
| RSF-COOH 1.0 | 221.1 | 38.39 | 22.54 | 191.7 | 195.7 | 43.30 | 61.01 |
| RSF-COOH 2.0 | 221.0 | 37.34 | 21.86 | 192.4 | 196.2 | 42.36 | 59.51 |

As can be seen in Table 2, the crystallization temperature gradually increased as the concentration of HBP increased. However, the crystallinity $X_c$ increased as the filler content increased to 1.0 phr, and then decreased when the content of HBP further increased. A similar trend was observed in ZhanG's study [47]. HBP can form hydrogen bonding with the PA 6 matrix and the filler surface, and the HBP molecules could act as the nucleating agent on the filler surface. When a low content of HBP was added to the composite (0–1.0 phr), crystallization of the PA 6 was accelerated by the increased amount of nuclei. As the HBP content was further increased beyond 1 phr, the growth of the crystal may be hindered by the excessive HBP molecules; as a result, the crystallinity was slightly reduced in RSF-COOH 2.0 compared to the RSF-COOH 1.0. The crystallization behavior of the RSF-COOH composites would affect the mechanical properties of the composites.

The crystallization kinetics of syntactic foams in Figure S7 and Table S2 showed that the crystallization rate was slowed by the further addition of HGM. The degree of crystallinity of the RSF composites calculated by the Avrami equation is depicted in Figure 4b. As demonstrated in Table 3, $t_{1/2}$ of the RSF-COOH composite decreased due to a higher crystallization rate constant as the content of HBP increased. HBP can be responsible for the crystallization rate, which was accelerated by not only enhanced chain mobility, but also by increased nucleation sites.

**Table 3.** Crystallization kinetic parameters of non-isothermal crystallization data for RSF containing various HBP contents.

| Sample Label | n | $Z_t$ | $Z_c$ | $t_{1/2}$ (min) | Adj. R-Square |
|---|---|---|---|---|---|
| RSF | 3.8 | 1.376 | 1.032 | 0.84 | 0.9997 |
| RSF-COOH 0.5 | 4.1 | 1.337 | 1.029 | 0.85 | 0.9997 |
| RSF-COOH 1.0 | 3.8 | 1.633 | 1.050 | 0.80 | 0.9997 |
| RSF-COOH 2.0 | 3.9 | 1.972 | 1.070 | 0.76 | 0.9997 |

## 3.4. Thermogravimetric Analysis

The extruded pellets were analyzed by TGA, which is presented in Figure 5. The residual char amount after the temperature reached 800 °C was 30 wt% in all composites, indicating the inorganic filler content in the RSF composites. The onset temperature of the thermal decomposition where the weight percent was 95% was increased from 408 to 415 °C when the HBP content increased from 0 to 0.75 phr. This result indicates that the thermal stability of the RSF composites was improved due to the enhanced interfacial adhesion between the filler and polymer matrix through the compatibilization which resulted from HBP [48]. When the HBP content was further increased to 2.0 phr, the decomposition occurred earlier at 403 °C. As a result of the lower thermal stability of HBP itself, which is presented in Figure S8, the excessive amount of HBP affects the thermal stability of the RSF-COOH composites where the content of HBP is higher than 1.0 phr.

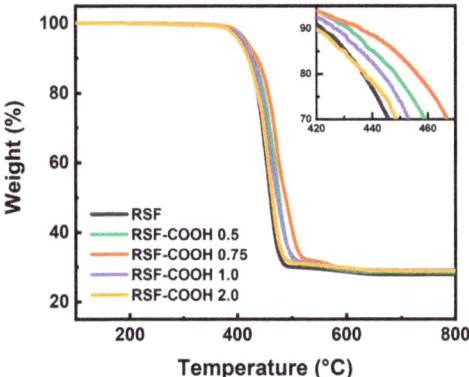

**Figure 5.** Thermogravimetric analysis data for RSF containing various HBP contents. Inset: A close-up of the 420–470 °C region.

## 3.5. Mechanical Properties and Specific Gravity of the Composite

Figure S9a and Table S3 show that increasing weight fraction of HGMs from 0 to 20 wt% did affect the mechanical properties of the composite significantly. When the weight fraction of HGM reached to 20 wt%, the tensile strength was significantly decreased by 40.6% due to the poor dispersion of the filler and the weak interfacial adhesion of the filler to the polymer matrix (Figure S4).

Figure 6 and Table 4 show the mechanical properties and the specific gravity of the PA 6, RSF control, and HBP containing RSF. The tensile strength increased by 24.3% in RSF-COOH 0.75 compared to the neat RSF. As a result of the abundant functional groups that can form hydrogen bonding, the strong interfacial adhesion between PA 6 polymers and fillers could enhance the tensile strength of the RSF-COOH composite. On the other hand, as the HBP content increased to 2.0 phr, tensile strength gradually decreased. It is interpreted that the physical property was weakened by the lubricating effect that comes from the dendritic structure of the HBP molecules. We can conclude that the optimal HBP content where the tensile strength is a maximum is 0.75–1.0 phr, and this indicates that the compatibility and the lubrication would be complementary at this content.

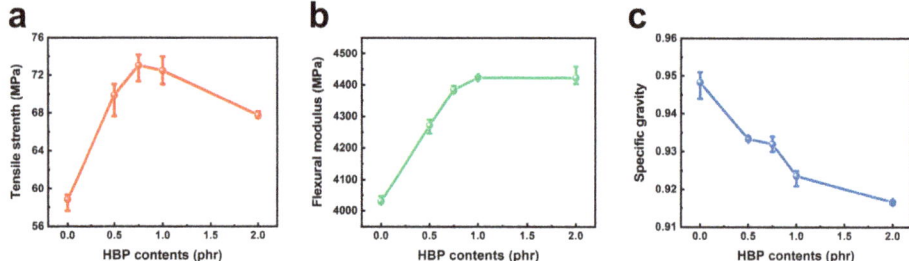

**Figure 6.** Tensile strength (**a**), flexural modulus (**b**), and the specific gravity (**c**) of the RSF with various contents of HBP.

**Table 4.** Comparison of the mechanical properties of neat PA 6 and RSF containing various HBP contents.

| Sample Code | Specific Gravity | Tensile Strength (MPa) | Elongation (%) | Flexural Modulus (MPa) | Flexural Strength (MPa) |
| --- | --- | --- | --- | --- | --- |
| PA 6 | 1.125 | 80.2 ± 2.6 | 13.6 ± 13.8 | 2843 ± 73 | 35.3 ± 0.9 |
| RSF | 0.948 | 58.8 ± 1.2 | 1.4 ± 0.2 | 4032 ± 16 | 50.0 ± 0.2 |
| RSF-COOH 0.5 | 0.933 | 69.9 ± 2.2 | 1.4 ± 0.6 | 4273 ± 26 | 53.0 ± 0.3 |
| RSF-COOH 0.75 | 0.932 | 73.1 ± 1.7 | 1.3 ± 0.3 | 4386 ± 13 | 54.6 ± 0.2 |
| RSF-COOH 1.0 | 0.924 | 72.5 ± 1.5 | 1.8 ± 0.6 | 4424 ± 0 | 54.9 ± 0 |
| RSF-COOH 2.0 | 0.917 | 70.7 ± 1.5 | 1.3 ± 0.2 | 4423 ± 36 | 54.6 ± 0.4 |

The flexural test results of the composite are summarized in Figure 6b and Table 4. HBP had a positive effect on the flexural modulus and flexural strength. In RSF-COOH 1.0, compared to RSF-control, the flexural modulus increased from 4032 to 4424 MPa, and the flexural strength increased from 50 to 55 MPa. The increase in flexural strength is a result of the high interfacial adhesion of the HBP molecules. Furthermore, on account of the reduced filler breakage during the processing, which is summarized in Table 5, foam-core sandwich structure of the RSF-COOH composites could be formed more effectively than that with the neat RSF.

**Table 5.** HGM breakage measurement results of each RSF composite containing various HBP contents.

| Sample Code | Ash Density [a] (g/cm$^3$) | Residual HGM Density (g/cm$^3$) | HGM Breakage (vol%) |
| --- | --- | --- | --- |
| RSF | 0.5829 | 0.4772 | 9.14 |
| RSF-COOH 0.5 | 0.5590 | 0.4596 | 6.07 |
| RSF-COOH 1.0 | 0.5463 | 0.4504 | 3.73 |
| RSF-COOH 2.0 | 0.5431 | 0.4470 | 2.84 |

[a] Measured by exposing the pellets to 550 °C for 2 h to remove the matrix (PA 6). The true density of the residual inorganic ashes consisting of HGMs and GFs was measured by gas pycnometer.

Specific gravity data of the RSF samples are provided in Figure 6c. It was shown that the specific gravity continued to decrease as the content of HBP increased. This is because the hollow HGMs can be damaged not only by the high shear in the extrusion process, but also by the additional injection molding. HBP brings out economic feasibility through tribological advantage, resulting in a much lighter RSF composite due to reduced HGM breakage. HGM breakage was calculated by TGA and gas pycnometer analysis in Table 5 and Figure S11. HGM breakage of the neat RSF and RSF-COOH 2.0 was 9.14 and 2.84 vol%, respectively, which show good agreement with the tendency of the specific gravity in Figure 6c. These results can be supplementary evidence that the remarkable variation in shear stress, mentioned in Figure 3c, improved the internal friction of the composite.

## 4. Conclusions

In this study, glass fiber-reinforced PA6 syntactic foam material was prepared with the addition of the carboxyl group-containing HBP for lightweight material applications. The morphological, rheological, and thermal properties of the composite material were characterized. It was revealed that HBP could increase mechanical properties, and could act as a rheological modifier simultaneously. The outcomes from the present study are summarized as follows:

(1) In the rheological characterization, the complex viscosity and the shear stress of the RSF-COOH composites decreased across the entire frequency range compared to the neat RSF composite without HBP. The storage modulus and loss modulus tended to decrease with increasing content of HBP in the composite; however, the slope of the storage modulus at low frequency tended to decrease. These results are evidence that the lubricating effect due to the dendritic structure of HBP, in addition to hydrogen bonding, were functioning in the RSF-COOH composite at the same time.

(2) The tensile strength and flexural modulus were increased by 24.3 and 9.7%, respectively, in RSF-COOH composite compared to the neat RSF. In the FE-SEM microphotographs, it was shown that the compatibility of the filler improved, and that filler-polymer adhesion enhanced when HBP was added. In addition, in DSC data, the crystallinity of the composite ($X_c$) increased to 22.54% with the addition of HBP in the composite. The enhanced compatibility of the filler with the polymer matrix, and crystallization behavior contributed to the increased mechanical strength of the RSF-COOH composite material.

(3) As a result of the reduced shear stress of the RSF-COOH composite in the high shear-applied extrusion process, HGM breakage considerably decreased from 9.14 to 2.84%. As a result, the specific gravity of the composite significantly decreased from 0.948 to 0.917 when HBP was introduced to the RSF composite.

The results show that HBP plays multiple roles in the RSF composite: modifying rheological properties, increasing mechanical properties, and reducing specific gravity. With these simultaneous effects in the RSF composite, HBP could be applied further in the lightweight materials field.

**Supplementary Materials:** The following are available online at https://www.mdpi.com/article/10.3390/polym14091915/s1: Figure S1: Twin-screw configuration representing extrusion process starting from left to right. HGMs and GFs are side-fed in the middle. Figure S2: DSC profile to figure out the effect of HBP contents within RSF samples on non-isothermal crystallization behavior. Data were collected from segment 6 on heating cycle to measure and calculate $T_m$, $\Delta H_f$, and $X_c$. Segment 8 was implemented on cooling cycle to measure and calculate $T_c$, $T_{onset}$, $\Delta H_c$, and $\Delta H_c^*$. Figure S3: FT-IR spectra of HBP. OH stretching in the carboxyl group is 2952 cm$^{-1}$, and aliphatic O-H bending is 1427 cm$^{-1}$. Aliphatic ester group of C-O stretching is 1191 cm$^{-1}$, and aliphatic ether group of C-O stretching is 1149 cm$^{-1}$. Figure S4: SEM microphotographs of syntactic foams on fractured surfaces after tensile test: PA 6/HGM 5 (a), PA 6/HGM 10 (b), PA 6/HGM 15 (c), and PA 6/HGM 20 (d). Figure S5: DSC thermograms of syntactic foams. The more the HGM was introduced, the lower the peak of crystallization temperature was found. Figure S6: DSC thermograms ranging from 20 to 100 °C of the PA 6 (a), RSF (b), RSF-COOH 0.5 (c), RSF-COOH 1.0 (d), and RSF-COOH 2.0 (e) for the glass transition temperature ($T_g$) evaluation. $T_g$ of each sample is depicted in the plots. Figure S7: Plots of relative crystallinity ($V_t$) vs time for the non-isothermal crystallization of neat PA 6, syntactic foams with different HGM contents, and reinforced PA 6 composites. Figure S8: TGA thermograms of PA 6 and HBP. Figure S9: Tensile strength (a), flexural modulus (b), and specific gravity (c) of syntactic foams with various HGM contents. Figure S10: Equations for calculating HGM breakage. The density of HGM reaches 2.54 g/cm$^3$ at 100% breakage. By exposing the pellets to 550 °C for 2 h to remove the matrix (PA 6), true densities of the residual inorganic ash consisting of HGMs and GFs were measured by gas pycnometer, which were 0.4365 and 2.4510 g/cm$^3$, respectively. Figure S11: TGA thermograms of syntactic foams and a glass fiber-reinforced PA 6 composite. We determined actual inorganic filler contents represented by the embedded table. Table S1: DSC data for syntactic foams and a glass fiber reinforced PA 6 composite under non-isothermal crystallization. Table S2:

Crystallization kinetic analysis of non-isothermal crystallization data for syntactic foams and a glass fiber-reinforced composite. Table S3: Comparison of the mechanical properties of neat PA 6, syntactic foams, and a glass fiber-reinforced PA 6 composite.

**Author Contributions:** Conceptualization, Y.Q.; methodology, J.K., J.L.; validation, K.P., S.H. (Sosan Hwang); formal analysis, investigation, and writing—original draft preparation, J.K., J.L.; writing—review and editing, Y.Q., S.E.S.; supervision, S.E.S., Y.Q.; project administration, S.H. (Sanghyun Hong), S.L.; funding acquisition, S.E.S. All authors have read and agreed to the published version of the manuscript.

**Funding:** This study was supported by LG Electronics, Co. Ltd.

**Institutional Review Board Statement:** Not applicable.

**Informed Consent Statement:** Not applicable.

**Data Availability Statement:** Not applicable.

**Conflicts of Interest:** Authors have no conflict to declare.

## References

1. Regulations for Emissions from Vehicles and Engines. Available online: https://www.epa.gov/regulations-emissions-vehicles-and-engines (accessed on 14 February 2022).
2. Sun, Q.; Zhou, G.; Meng, Z.; Jain, M.; Su, X. An integrated computational materials engineering framework to analyze the failure behaviors of carbon fiber reinforced polymer composites for lightweight vehicle applications. *Compos. Sci. Technol.* **2021**, *202*, 108560. [CrossRef] [PubMed]
3. Mallick, P.K. *Materials, design and manufacturing for lightweight vehicles*; Woodhead Publishing: Sawston, UK, 2020.
4. Graziano, A.; Garcia, C.; Jaffer, S.; Tjong, J.; Yang, W.; Sain, M. Functionally tuned nanolayered graphene as reinforcement of polyethylene nanocomposites for lightweight transportation industry. *Carbon* **2020**, *169*, 99–110. [CrossRef]
5. Du, Y.; Keller, T.; Song, C.; Wu, L.; Xiong, J. Origami-inspired carbon fiber-reinforced composite sandwich materials—Fabrication and mechanical behavior. *Compos. Sci. Technol.* **2021**, *205*, 108667. [CrossRef]
6. Ozden, S.; Dutta, N.S.; Randazzo, K.; Tsafack, T.; Arnold, C.B.; Priestley, R.D. Interfacial engineering to tailor the properties of multifunctional ultralight weight hBN-polymer composite aerogels. *ACS Appl. Mater. Interf.* **2021**, *13*, 13620–13628. [CrossRef] [PubMed]
7. Choi, D.; Kil, H.-S.; Lee, S. Fabrication of low-cost carbon fibers using economical precursors and advanced processing technologies. *Carbon* **2019**, *142*, 610–649. [CrossRef]
8. Zhang, X.; Wang, P.; Zhou, Y.; Li, X.; Yang, E.-H.; Yu, T.X.; Yang, J. The effect of strain rate and filler volume fraction on the mechanical properties of hollow glass microsphere modified polymer. *Compos. Part B Eng.* **2016**, *101*, 53–63. [CrossRef]
9. Cosse, R.L.; Araújo, F.H.; Pinto, F.A.N.C.; de Carvalho, L.H.; de Morais, A.C.L.; Barbosa, R.; Alves, T.S. Effects of the type of processing on thermal, morphological and acoustic properties of syntactic foams. *Compos. Part B Eng.* **2019**, *173*, 106933. [CrossRef]
10. Jayavardhan, M.L.; Doddamani, M. Quasi-static compressive response of compression molded glass microballoon/HDPE syntactic foam. *Compos. Part B Eng.* **2018**, *149*, 165–177. [CrossRef]
11. Jayavardhan, M.L.; Bharath Kumar, B.R.; Doddamani, M.; Singh, A.K.; Zeltmann, S.E.; Gupta, N. Development of glass microballoon/HDPE syntactic foams by compression molding. *Compos. Part B Eng.* **2017**, *130*, 119–131. [CrossRef]
12. Zhang, L.; Ma, J. Effect of coupling agent on mechanical properties of hollow carbon microsphere/phenolic resin syntactic foam. *Compos. Sci. Technol.* **2010**, *70*, 1265–1271. [CrossRef]
13. Tagliavia, G.; Porfiri, M.; Gupta, N. Analysis of flexural properties of hollow-particle filled composites. *Compos. Part B Eng.* **2010**, *41*, 86–93. [CrossRef]
14. Wang, L.; Yang, X.; Zhang, J.; Zhang, C.; He, L. The compressive properties of expandable microspheres/epoxy foams. *Compos. Part B Eng.* **2014**, *56*, 724–732. [CrossRef]
15. Gupta, N.; Zeltmann, S.E.; Shunmugasamy, V.C.; Pinisetty, D. Applications of polymer matrix syntactic foams. *Jom* **2013**, *66*, 245–254. [CrossRef]
16. Swetha, C.; Kumar, R. Quasi-static uni-axial compression behaviour of hollow glass microspheres/epoxy based syntactic foams. *Mater. Des.* **2011**, *32*, 4152–4163. [CrossRef]
17. He, S.; Carolan, D.; Fergusson, A.; Taylor, A.C. Toughening epoxy syntactic foams with milled carbon fibres: Mechanical properties and toughening mechanisms. *Mater. Des.* **2019**, *169*, 107654. [CrossRef]
18. Gogoi, R.; Manik, G.; Arun, B. High specific strength hybrid polypropylene composites using carbon fibre and hollow glass microspheres: Development, characterization and comparison with empirical models. *Compos. Part B Eng.* **2019**, *173*, 106875. [CrossRef]

19. Kumar, N.; Mireja, S.; Khandelwal, V.; Arun, B.; Manik, G. Light-weight high-strength hollow glass microspheres and bamboo fiber based hybrid polypropylene composite: A strength analysis and morphological study. *Compos. Part B Eng.* **2017**, *109*, 277–285. [CrossRef]
20. Hu, Y.; Mei, R.; An, Z.; Zhang, J. Silicon rubber/hollow glass microsphere composites: Influence of broken hollow glass microsphere on mechanical and thermal insulation property. *Compos. Sci. Technol.* **2013**, *79*, 64–69. [CrossRef]
21. Lin, C.-A.; Ku, T.-H. Shear and elongational flow properties of thermoplastic polyvinyl alcohol melts with different plasticizer contents and degrees of polymerization. *J. Mater. Process. Technol.* **2008**, *200*, 331–338. [CrossRef]
22. Wu, C.; McGinity, J.W. Influence of methylparaben as a solid-state plasticizer on the physicochemical properties of Eudragit® RS PO hot-melt extrudates. *Eur. J. Pharm. Biopharm.* **2003**, *56*, 95–100. [CrossRef]
23. Belous, A.; Tchoudakov, R.; Tzur, A.; Narkis, M.; Alperstein, D. Development and characterization of plasticized polyamides by fluid and solid plasticizers. *Polym. Adv. Technol.* **2012**, *23*, 938–945. [CrossRef]
24. Wang, J.; Ye, Z.; Zhu, S. Topology-engineered hyperbranched high-molecular-weight polyethylenes as lubricant viscosity-index improvers of high shear stability. *Ind. Eng. Chem. Res.* **2007**, *46*, 1174–1178. [CrossRef]
25. Wang, F.C.-Y.; Buzanowski, W.C. Polymer additive analysis by pyrolysis–gas chromatography. *J. Chromatogr. A* **2000**, *891*, 313–324. [CrossRef]
26. Bettini, S.H.; de Miranda Josefovich, M.P.; Munoz, P.A.; Lotti, C.; Mattoso, L.H. Effect of lubricant on mechanical and rheological properties of compatibilized PP/sawdust composites. *Carbohydr. Polym.* **2013**, *94*, 800–806. [CrossRef]
27. Jacobsen, S.; Fritz, H.-G. Plasticizing polylactide—the effect of different plasticizers on the mechanical properties. *Polym. Eng. Sci.* **1999**, *39*, 1303–1310. [CrossRef]
28. Cosimbescu, L.; Robinson, J.W.; Zhou, Y.; Qu, J. Dual functional star polymers for lubricants. *RSC Adv.* **2016**, *6*, 86259–86268. [CrossRef]
29. Robinson, J.W.; Zhou, Y.; Bhattacharya, P.; Erck, R.; Qu, J.; Bays, J.T.; Cosimbescu, L. Probing the molecular design of hyperbranched aryl polyesters towards lubricant applications. *Sci. Rep.* **2016**, *6*, 18624. [CrossRef]
30. Zheng, Y.; Li, S.; Weng, Z.; Gao, C. Hyperbranched polymers: Advances from synthesis to applications. *Chem. Soc. Rev.* **2015**, *44*, 4091–4130. [CrossRef]
31. Hawker, C.J.; Farrington, P.J.; Mackay, M.E.; Wooley, K.L.; Frechet, J.M. Molecular ball bearings: The unusual melt viscosity behavior of dendritic macromolecules. *J. Am. Chem. Soc.* **1995**, *117*, 4409–4410. [CrossRef]
32. Chen, L.; Qin, Y.; Wang, X.; Li, Y.; Zhao, X.; Wang, F. Toughening of poly(propylene carbonate) by hyperbranched poly(ester-amide) via hydrogen bonding interaction. *Polym. Inter.* **2011**, *60*, 1697–1704. [CrossRef]
33. Wu, C.; Huang, X.; Wang, G.; Wu, X.; Yang, K.; Li, S.; Jiang, P. Hyperbranched-polymer functionalization of graphene sheets for enhanced mechanical and dielectric properties of polyurethane composites. *J. Mater. Chem.* **2012**, *22*, 7010–7019. [CrossRef]
34. Gu, W.; Liu, X.; Gao, Q.; Gong, S.; Li, J.; Shi, S.Q. Multiple hydrogen bonding enables strong, tough, and recyclable soy protein films. *ACS Sustain. Chem. Eng.* **2020**, *8*, 7680–7689. [CrossRef]
35. Peng, Q.; Li, Y.; He, X.; Lv, H.; Hu, P.; Shang, Y.; Wang, C.; Wang, R.; Sritharan, T.; Du, S. Interfacial enhancement of carbon fiber composites by poly(amido amine) functionalization. *Compos. Sci. Technol.* **2013**, *74*, 37–42. [CrossRef]
36. Avrami, M. Granulation, Phase change, and microstructure kinetics of phase change. III. *J. Chem. Phy.* **1941**, *9*, 177–184. [CrossRef]
37. Avrami, M. Kinetics of phase change. I General theory. *J. Chem. Phy.* **1939**, *7*, 1103–1112. [CrossRef]
38. Avrami, M. Kinetics of phase change. II Transformation-Time relations for random distribution of nuclei. *J. Chem. Phy.* **1940**, *8*, 212–224. [CrossRef]
39. Fornes, T.D.; Paul, D.R. Crystallization behavior of nylon 6 nanocomposites. *Polymer* **2003**, *44*, 3945–3961. [CrossRef]
40. Fan, Z.; Jaehnichen, K.; Desbois, P.; Haeussler, L.; Vogel, R.; Voit, B. Blends of different linear polyamides with hyperbranched aromatic AB2 and A2+ B3 polyesters. *J. Polym. Sci. Part A Polym. Chem.* **2009**, *47*, 3558–3572. [CrossRef]
41. Kang, D.; Hwang, S.W.; Jung, B.N.; Shim, J.K. Effect of hollow glass microsphere (HGM) on the dispersion state of single-walled carbon nanotube (SWNT). *Compos. Part B Eng.* **2017**, *117*, 35–42. [CrossRef]
42. Goldansaz, H.; Goharpey, F.; Afshar-Taromi, F.; Kim, I.; Stadler, F.J.; van Ruymbeke, E.; Karimkhani, V. Anomalous Rheological Behavior of Dendritic Nanoparticle/Linear Polymer Nanocomposites. *Macromolecules* **2015**, *48*, 3368–3375. [CrossRef]
43. Bhardwaj, R.; Mohanty, A.K. Modification of brittle polylactide by novel hyperbranched polymer-based nanostructures. *Biomacromolecules* **2007**, *8*, 2476–2484. [CrossRef] [PubMed]
44. Carolan, D.; Mayall, A.; Dear, J.P.; Fergusson, A.D. Micromechanical modelling of syntactic foam. *Compos. Part B Eng.* **2020**, *183*, 107701. [CrossRef]
45. Privalko, V.; Kawai, T.; Lipalov, Y.S. Crystallization of filled nylon 6 III. Non-isothermal crystallization. *Colloid Polym. Sci.* **1979**, *257*, 1042–1048. [CrossRef]
46. Jang, K.-S. Low-density polycarbonate composites with robust hollow glass microspheres by tailorable processing variables. *Polym. Test.* **2020**, *84*, 106408. [CrossRef]
47. Zhang, J.-F.; Sun, X. Mechanical properties and crystallization behavior of poly(lactic acid) blended with dendritic hyperbranched polymer. *Polym. Inter.* **2004**, *53*, 716–722. [CrossRef]
48. Lu, S.; Li, S.; Yu, J.; Guo, D.; Ling, R.; Huang, B. The effect of hyperbranched polymer lubricant as a compatibilizer on the structure and properties of lignin/polypropylene composites. *Wood Mater. Sci. Eng.* **2013**, *8*, 159–165. [CrossRef]

Article

# Thermal and Adhesion Properties of Fluorosilicone Adhesives Following Incorporation of Magnesium Oxide and Boron Nitride of Different Sizes and Shapes

Kyung-Soo Sung [1], So-Yeon Kim [1], Min-Keun Oh [2] and Namil Kim [2,*]

1. Research & Development Center, Protavic Korea, Daejeon 34326, Korea; kssung@protavic.co.kr (K.-S.S.); sykim@protavic.co.kr (S.-Y.K.)
2. Energy Materials R&D Center, Korea Automotive Technology Institute, Cheonan-si 31214, Korea; mkoh@katech.re.kr
* Correspondence: nikim@katech.re.kr; Tel.: +82-41-559-3241

**Abstract:** Thermally conductive adhesives were prepared by incorporating magnesium oxide (MgO) and boron nitride (BN) into fluorosilicone resins. The effects of filler type, size, and shape on thermal conductivity and adhesion properties were analyzed. Higher thermal conductivity was achieved when larger fillers were used, but smaller ones were advantageous in terms of adhesion strength. Bimodal adhesives containing spherical MgOs with an average particle size of 120 μm and 90 μm exhibited the highest conductivity value of up to 1.82 W/mK. Filler shape was also important to improve the thermal conductivity as the filler type increased. Trimodal adhesives revealed high adhesion strength compared to unimodal and bimodal adhesives, which remained high after aging at 85 °C and 85% relative humidity for 168 h. It was found that the thermal and adhesion properties of fluorosilicone composites were strongly affected by the packing efficiency and interfacial resistance of the particles.

**Keywords:** adhesive; fluorosilicone; thermal conductivity; magnesium oxide; boron nitride

## 1. Introduction

As electronic devices become increasingly smaller and more highly integrated, effective heat dissipation in microelectronic packaging has become a critical issue to ensure their reliability [1–3]. Recently, the wide use of power semiconductors has accelerated the necessity of faster heat transfer materials along with long-term thermal stability above 200 °C. A small difference in operating temperature can lead to working instability and reduction of the life span of devices. Thermally conductive adhesives (TCAs) are used as interconnected materials for the purpose of heat dissipation from chips and mechanical support by suitable adhesion with metal substrates. As the application area of TCAs has been expanded, a variety of properties such as thermal stability, chemical resistance, and low thermal expansion have additionally become required [4–6].

Epoxy resins have been widely used as adhesives and coating solutions because of their excellent adhesion strength, low shrinkage, and ease of curing [7–10]. However, they intrinsically possess poor fracture resistance due to limited flexibility and low thermal stability. To overcome these problems, impact modifiers including rubber and polyurethane silicone are blended [11–13]. Silicone resins have attracted interest in pressure-sensitive adhesives and electronics packaging because of their excellent thermal stability, good processability, and hydrophobicity [14–16]. Since most silicone resins have poor solvent resistance, they are often modified with fluorine groups. Fluorosilicone resins combine the structures of fluorocarbon and polysiloxane. The unique properties of heat resistance, low-temperature flexibility, and chemical resistance make them suitable for high-value products in the electronics industry [17–19]. Despite such advantages, the characteristics of fluorosilicone adhesives are not fully understood due to their low mechanical strength and high cost.

Ceramic fillers are commonly used in adhesive composites to provide thermal conductivity while maintaining electrical insulation and dielectric breakdown voltage [20–23]. High thermal conductivity of organic resins can be achieved by forming large numbers of conductive pathways and reducing the interfacial resistance between the fillers and the resins. Ceramic fillers with a high aspect ratio or larger size have less filler-polymer interfaces and, therefore, lower phonon scattering. Although the thermal conductivity of adhesives is mainly affected by the inherent properties of fillers, higher filler loading is unavoidable in order to meet the required thermal conductivity in electronic devices. High filler loading generally accompanies poor processability due to high viscosity and the formation of voids between the fillers, which are detrimental to the thermal conductivity. Therefore, the filler content in adhesives is rather limited. Hybrid adhesives containing two or more types of fillers with different shapes and sizes have been suggested to attain high thermal conductivity and good processability [24–27]. Since a combination of conductive fillers can enhance the filling density in which small particles can occupy the space between adjacent large particles, thermally conductive networks with low thermal resistance can be formed at reduced filler content.

In the present study, we compared the thermal and chemical properties of adhesive resins using epoxy, silicone, and fluorosilicone resins. Magnesium oxide (MgO) and boron nitride (BN) were added to offer thermal conductivity. MgO has many attractive characteristics, such as a bulk thermal conductivity greater than alumina, and moreover, it is inexpensive and nontoxic [28,29]. Meanwhile, BN is promising to obtain high thermal conductive adhesives with excellent electric insulation [30,31]. Approximated properties of ceramic fillers and their estimated recent costs are summarized in Table 1. The effect of filler type, size, and shape on thermal conductivity and adhesion strength was investigated. Hybrid adhesives containing more than two types of ceramic fillers were fabricated to acquire better thermal conductivity and adhesion strength at the same content. The reliability of the adhesives was proved by monitoring the change of adhesion strength after aging in high thermal (85 °C) and humidity (85%) environments.

Table 1. Comparison of intrinsic properties and estimated cost of commercial ceramic fillers.

| Ceramic Filler | Mohs Hardness, 20 °C (Hv) | Thermal Conductivity, 20 °C (W/mK) | Specific Gravity | Coefficient of Thermal Expansion, 0~1000 °C ($\times 10^6$/°C) | Dielectric Constant, 20 °C, 1 MHz | Estimated Cost, 2021 ($/kg) |
|---|---|---|---|---|---|---|
| Aluminum Oxide ($Al_2O_3$) | 9 | 32 | 3.9 | 8 | 8.9 | 20–30 |
| Aluminum Nitride (AlN) | 8 | 320 | 3.3 | 5.6 | 8.8 | 100–150 |
| Hexagoal Boron Nitride (h-BN) | 2 | 275 | 2.3 | 2.8 | 4.5 | 100–150 |
| Silicon Carbide (SiC) | 10 | 190 | 3.2 | 4.5 | 11 | 100–150 |
| Silicon Nitride ($Si_3N_4$) | 9 | 26 | 3.2 | 3.5 | 8 | 30–40 |
| Magnesium Oxide (MgO) | 5 | 60 | 3.6 | 11 | 12 | 30–40 |
| Fused Silicon Dioxide ($SiO_2$) | 3.5 | 1 | 2.6 | 0.5 | 3.8 | 5–10 |

## 2. Materials and Methods

### 2.1. Materials and Sample Preparation

The epoxy resins were composed of 60.0 wt% cycloaliphatic epoxy (C-2021P, Daicel, Tokyo, Japan), 39.0 wt% curing agent (RIKACID MH, New Japan Chemical, Osaka, Japan), 0.5 wt% accelerator (2E4MZ-CN, Shikoku Chemicals, Kagawa, Japan), and 0.5 wt% adhesion promoter (KBM403, Shin-Etsu, Tokyo, Japan). The Silicone resins were prepared from 85 wt% vinyl silicone (Andisil VS 1000, AB Specialty Silicones, Waukegan, IL, USA),

8.0 wt% and 5.9 wt% crosslinkers (Andisil XL1341 and XL1342, AB Specialty Silicones, Waukegan, IL, USA), 0.1 wt% platinum catalyst (C1142A, Johnson Matthey, London, UK), and 1 wt% adhesion promoter (KBM 1003, Shin-Etsu, Tokyo, Japan). The Fluorosilicone resins contained 53.8 wt% of vinyl fluorosilicone (FS-8019-1000, TOPDA, Fuzhou, China), 45.0 wt% fluoro crosslinker (FS-8016, TOPDA, Fuzhou, China), 0.2 wt% platinum catalyst (C1142A, Johnson Mattey, London, UK), and 1.0 wt% adhesion promoter (KBM 1003, Shin-Etsu, Tokyo, Japan). The mixtures of respective resin were prepared at room temperature by stirring mechanically for 1 min, followed by the three-roll mill technique (EXAKT 80E, EXAKT, Norderstedt, Germany) until the solution became completely homogeneous. All the resins were thermally cured at 150 °C for 60 min.

Five types of magnesium oxides (MgOs) with spherical and amorphous shapes were used as conductive fillers. Three spherical MgOs with an average particle size of 120 μm, 90 μm, and 50 μm were purchased from Denka Corp. Ltd. (Tokyo, Japan) while two amorphous MgOs with an average particle size of 6 μm and 0.6 μm were obtained from Konoshima Chemical Co., Ltd. (Osaka, Japan). Platelet-shaped boron nitrides (BNs) bought from Denka Co. Ltd. (Tokyo, Japan), had an average particle size of 5 μm and 4 μm. Aggregated and flake BNs with an average particle size of 12 μm and 280 μm were acquired from Denka and 3M Co., Ltd. (Kempten, Germany), respectively. A predetermined ratio of MgO and BN was uniformly dispersed into the fluorosilicone resins using a conventional ceramic type three roll mill and then rotated at 100 rpm for 30 min in a vacuum bath to remove the residual bubbles. The mixtures were stored in a refrigerator at −40 °C prior to use. Thermal curing was performed at 150 °C for 60 min. The composition of the bimodal and trimodal adhesives is listed in Table 2.

**Table 2.** Composition of bimodal and trimodal adhesives containing MgO and BN with various sizes and shapes.

| Composition | MgO (vol%) | | | | | BN (vol%) | | | |
| --- | --- | --- | --- | --- | --- | --- | --- | --- | --- |
| | 120 μm (Sphere) | 90 μm (Sphere) | 50 μm (Sphere) | 6 μm (Amorp) | 0.6 μm (Amorp) | 5 μm (Plate) | 4 μm (Plate) | 12 μm (Aggreg) | 280 μm (Flake) |
| Bi-MgO-1 | 45 | 15 | / | / | / | / | / | / | / |
| Bi-MgO-2 | 45 | / | 15 | / | / | / | / | / | / |
| Bi-MgO-3 | 45 | / | / | 15 | / | / | / | / | / |
| Bi-MgO-4 | 45 | / | / | / | 15 | / | / | / | / |
| Bi-MgO/BN-1 | 45 | / | / | / | / | 15 | / | / | / |
| Bi-MgO/BN-2 | 45 | / | / | / | / | / | 15 | / | / |
| Bi-MgO/BN-3 | 45 | / | / | / | / | / | / | 15 | / |
| Bi-MgO/BN-4 | 45 | / | / | / | / | / | / | / | 15 |
| Tri-MgO-1 | 40 | 13.3 | 6.7 | / | / | / | / | / | / |
| Tri-MgO-2 | 40 | 13.3 | / | 6.7 | / | / | / | / | / |
| Tri-MgO-3 | 40 | 13.3 | / | / | 6.7 | / | / | / | / |
| Tri-MgO2/BN1-1 | 40 | 13.3 | / | / | / | 6.7 | / | / | / |
| Tri-MgO2/BN1-2 | 40 | 13.3 | / | / | / | / | 6.7 | / | / |
| Tri-MgO2/BN1-3 | 40 | 13.3 | / | / | / | / | / | 6.7 | / |
| Tri-MgO2/BN1-4 | 40 | 13.3 | / | / | / | / | / | / | 6.7 |
| Tri-MgO1/BN2-1 | 40 | / | / | / | / | 6.7 | / | / | 13.3 |
| Tri-MgO1/BN2-2 | 40 | / | / | / | / | / | 6.7 | / | 13.3 |
| Tri-MgO1/BN2-3 | 40 | / | / | / | / | / | / | 6.7 | 13.3 |

## 2.2. Characterization

The shear force required to detach a square silicon die (4 mm × 4 mm, 0.35 mm of thickness) from a copper lead frame (10 mm × 10 mm, 0.2 mm of thickness) was measured to evaluate the adhesion strength using a Dage Series 4000 micro-tester (Dage, Aylesbury, UK) at a movement speed of 2 mm/s. The solution weighing around 10 mg was spread on a lead fame and then a die was placed on it. The adhesive thickness was about 200 μm. The hot die shear strength was further tested at 180 °C and 250 °C, which corresponded to the wire bonding and reflow temperature, respectively. The results were averaged from five sets of tested specimens. The effect of heat and moisture exposure on the adhesion strength was examined by placing the respective specimens in a temperature and humidity chamber (TH-ME-025, Jeiotech, Seoul, Korea) maintained at 85 °C and 85% relative humidity (RH). A fully cured epoxy, silicone, and fluorosilicone film with a dimension of 10 mm in width × 10 mm in length × 1.5 mm in thickness was immersed in various solvents with different polarities at room temperature for 168 h and then the mass changes were measured as a function of immersion time. Thermal conductivity measurement of the adhesive composites was carried out by the transient plane source (TPS) technique (TPS 500S, HotDisk, Göteborg, Sweden). The samples were prepared by depositing the adhesive solution onto Teflon plate covered with aluminum foil having a square well with a length of 20 mm and 0.5 mm in thickness. After thermal curing, the composites were removed from the plate. The morphological appearance of the fractured surface was observed by a scanning electron microscope (SEM) (SIGMA500, ZEISS, Jena, Germany). The specimens were sputtered with silver for 90 s using a sputtering device before characterization.

## 3. Results

### 3.1. Characteristics of Epoxy, Silicone, and Fluorosilicone Resins

The thermal property of epoxy, silicone, and fluorosilicone resins was estimated by investigating adhesion strength at 23 °C, 180 °C, and 250 °C. At 23 °C, the epoxy revealed an outstanding adhesion strength of 120.5 kgf/cm$^2$, which was more than two and fourteen times higher than silicone and fluorosilicone, respectively (Figure 1a). When the measurement was conducted at 180 °C, the adhesion strength of epoxy decreased drastically to 34.2 kgf/cm$^2$, while the silicone and fluorosilcone maintained their initial strength at 23 °C. At 250 °C, the adhesion strength of the resins was reduced to 9.5 kgf/cm$^2$ for epoxy, 14.7 kgf/cm$^2$ for silicone, and 6.7 kgf/cm$^2$ for fluorosilicone. Although the adhesion strength of fluorosilicone was lower than that of epoxy and silicone over the whole temperature range measured, they retained a high percentage of their room temperature properties at elevated temperatures. The adhesion test was conducted after humid and thermal aging at 85 °C and 85% RH for 168 h. As shown in Figure 1b, the adhesion strength of epoxy and silicone decreased from 120.5 kgf/cm$^2$ to 91.3 kgf/cm$^2$ and 43.9 kgf/cm$^2$ to 35.8 kgf/cm$^2$, respectively. Epoxy exhibited the lowest adhesion strength below 1 kgf/cm$^2$ at 250 °C, implying that epoxy is not suitable for lengthy exposure to high temperature and humid environments. On the other hand, fluorosilicone showed a slight increase from 8.4 kgf/cm$^2$ to 11.6 kgf/cm$^2$ at 23 °C, probably due to additional cross-linking reactions. Adhesion strength gradually decreased with increasing temperature. The adhesives in electronic circuits are commonly exposed to high-temperature environments and, therefore, the preservation of adhesion strength is an important requirement. Figure 1c displays the change of adhesion strength as a function of aging time at 250 °C. Although epoxy and silicone showed a relatively high strength of 120.5 kgf/cm$^2$ and 43.9 kgf/cm$^2$, the degree of reduction was less than 25% of the initial strength after 30 h. In contrast to neat epoxy and silicone, fluorosilicone retained a high percentage of its initial strength with exposure to high temperatures. The adhesion strength persisted for 20 h and then slightly decreased by about 25% after 30 h.

The degree of swelling was measured by immersing the cured resins in various solvents with different polarities. The polarity of the solvents used in the test decreased in the following sequence: acetone > tetrahydrofuran > toluene > normal hexane. As illustrated in Table 3, the polar solvents promoted a higher swelling tendency for all resins. Although a moderate swelling ratio of 30.4% and 15.2% were observed in tetrahydrofuran and acetone after 168 h, the swelling ratio of epoxy resins in n-hexane and toluene was less than 5%, probably due to higher cross-linking density. On the other hand, silicone resins showed an excessive swelling ratio above 50% except for in acetone, which may limit its electronics usage. With the incorporation of a fluorine group with silicone, the swelling ratio was suppressed drastically to below 20% in n-hexane and toluene. The increase in polarity in the presence of a fluorine group led to an increase in the swelling ratio in tetrahydrofuran and acetone due to pronounced affinity. From the thermal aging and swelling tests, it is apparent that the fluorosilicone resins possess promising characteristics such as chemical, heat, and humidity resistance. The thermal conductivity behavior of fluorosilicone adhesives was further analyzed by incorporating MgO and BN fillers.

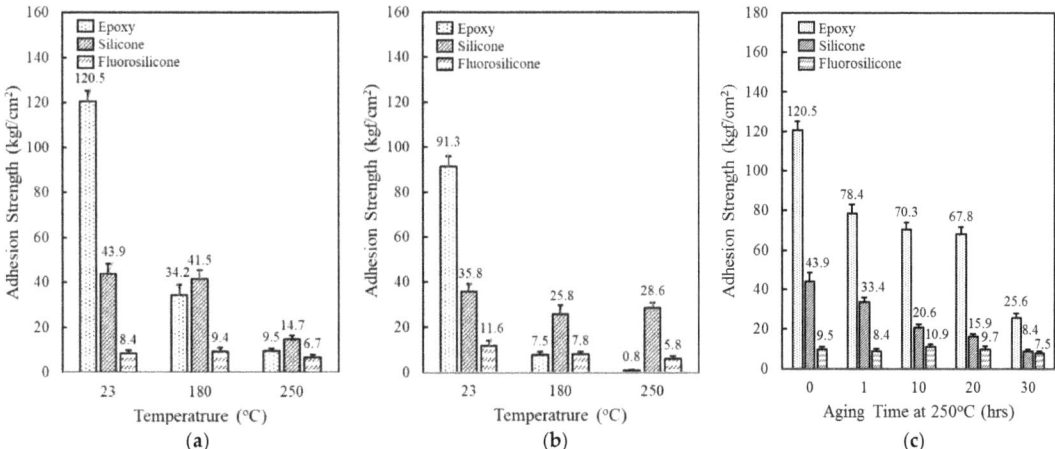

**Figure 1.** Comparison of adhesion strength of epoxy, silicone, and fluorosilicone resins: (**a**) Adhesion strength measured at 23 °C, 180 °C, 250 °C; (**b**) Adhesion strength measured at 23 °C, 180 °C, 250 °C after aging at 85 °C/85% RH for 168 h; (**c**) Adhesion strength as a function of aging time at 250 °C.

**Table 3.** The swelling ratio of epoxy, silicone, and fluorosilicone resins in various solvents.

| Resin | Swelling Ratio (%) with Immersion Time (h) in Solvent | | | | | | | | | | | | | | | |
|---|---|---|---|---|---|---|---|---|---|---|---|---|---|---|---|---|
| | n-Hexane | | | | Toluene | | | | Tetrahydrofuran | | | | Acetone | | | |
| | 24 | 48 | 72 | 168 | 24 | 48 | 72 | 168 | 24 | 48 | 72 | 168 | 24 | 48 | 72 | 168 |
| Epoxy | 1.2 | 2.9 | 3.3 | 4.8 | 0.0 | 0.8 | 0.4 | 3.5 | 6.2 | 7.5 | 22.5 | 30.4 | 0.9 | 7.0 | 10.8 | 15.2 |
| Silicone | 54.5 | 56.0 | 58.4 | 59.9 | 44.9 | 46.2 | 49.3 | 52.6 | 55.3 | 56.6 | 49.3 | 57.4 | 11.8 | 18.2 | 19.3 | 24.9 |
| Fluorosilicone | 8.5 | 12.3 | 13.6 | 15.0 | 15.6 | 15.7 | 15.4 | 18.1 | 24.9 | 27.0 | 27.4 | 29.4 | 25.3 | 23.6 | 24.1 | 26.2 |

*3.2. Thermal and Mechanical Properties of Unimodal Adhesives*

Figure 2a shows the variation of thermal conductivity of fluorosilicone adhesives containing MgO and BN of different sizes and shapes. The filler concentration was kept at 60 vol%. When MgO were incorporated, higher thermal conductivity was obtained using larger fillers. The thermal conductivity value of adhesives filled with 120 μm MgO was more than two times higher than 0.6 μm MgO, i.e., 1.67 W/mK for 120 μm and 0.73 W/mK for 0.6 μm. Small fillers have a large interfacial contact area with the resins and thus more phonon scattering occurs. In addition, spherical MgOs have favorable filler packing for facile heat dissipation through the composites. As adhesives in electronic circuits not only provide heat dissipation but also mechanical support, adhesion strength is another critical parameter. As shown in Figure 2b, the use of smaller fillers was favorable in order to obtain high adhesion strength. The adhesives filled with 0.6 μm MgO had an adhesion strength value of 14.7 kgf/cm$^2$ at 23 °C, while the adhesives with 120 μm MgO had a strength value of 12.0 kgf/cm$^2$. The adhesion strength gradually decreased as the test temperature increased. For example, an adhesion strength of 0.6μm MgO-filled adhesives was reduced from 14.7 kgf/cm$^2$ to 12.8 kgf/cm$^2$ at 180 °C and 11.3 kgf/cm$^2$ at 250 °C, respectively.

The reduced adhesion strength may be attributable to the difference in thermal expansion between adhesives and Cu substrates.

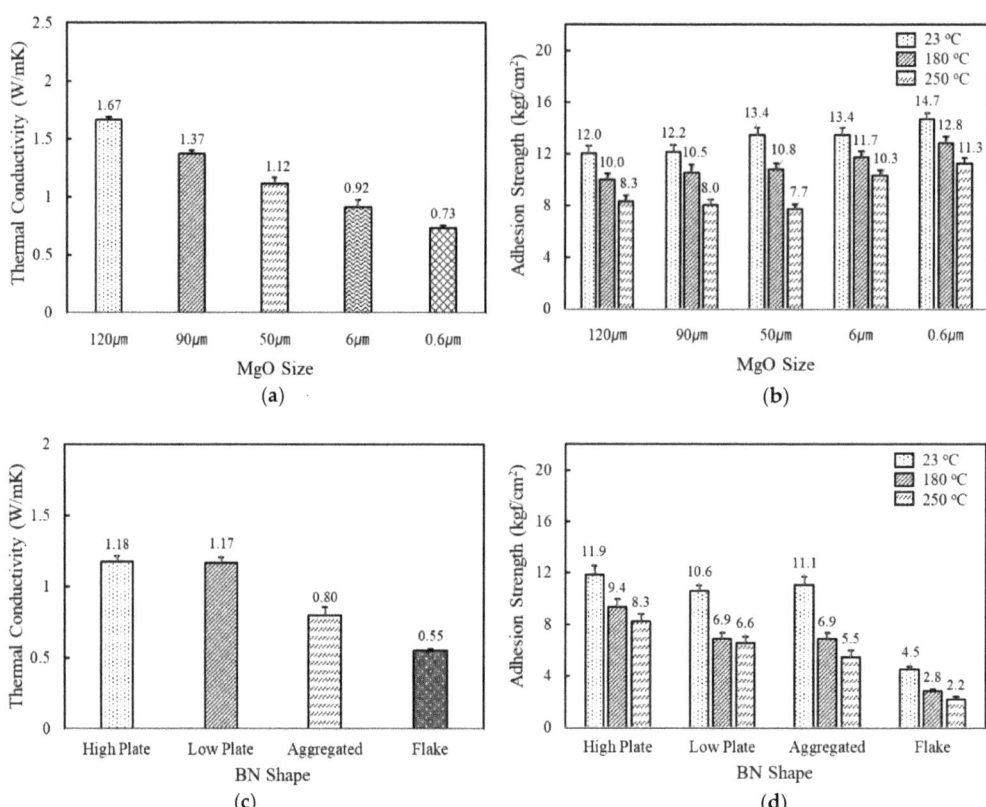

**Figure 2.** The thermal conductivity and adhesion strength of fluorosilicone adhesives filled with MgOs and BNs: (**a**) thermal conductivity of MgO-filled fluorosilicone; (**b**) adhesion strength of MgO-filled fluorosilicone measured at 23 °C, 180 °C, and 250 °C; (**c**) thermal conductivity of BN-filled fluorosilicone; (**d**) adhesion strength of BN-filled fluorosilicone measured at 23 °C, 180 °C, and 250 °C. The filler concentration was kept at 60 vol%.

When BNs are added to fluorosilicone resins instead of MgOs, the adhesives showed a lower thermal conductivity in a range of 0.55~1.18 W/mK at the same volume fraction. Platelet BNs with different crystallinity revealed a similar thermal conductivity value of 1.18 W/mK for high crystallinity and 1.17 W/mK for low crystallinity, indicating that the effect of filler crystallinity on thermal conductivity was negligible. In general, BNs in composites are known to exhibit a preferential alignment along the in-plane axis, resulting in large in-plane thermal conductivity. The in-plane thermal conductivity was more than 20 times higher than the through-plane value [32]. In our system, the thermal conductivity was measured by a transient plane source (TPS) method, where the amount of heat per unit time and unit area through a plate of unit thickness was measured. Therefore, the bulk thermal conductivity of BN-filled composites was expected to be low. The thermal conductivity of adhesives filled with aggregated and flake shaped BNs having an average particle size of 12 μm and 280 μm was observed below 1.0 W/mK. The adhesives containing flake shaped BNs underwent phase separation at 60 vol%, while aggregated BNs possessing a low specific surface area formed relatively low network concentrations. The adhesion strength of the BN-filled adhesives was found to be low compared to MgOs at below 12.0 kgf/cm$^2$ at 23 °C. The flake shaped BNs showed the lowest adhesion strength due to the large particle size and phase separation. On the basis of the thermal conductivity and adhesion strength results, it was found that the thermal and mechanical properties of unimodal adhesives mainly rely on the filler size and type.

Figure 3a shows the variation of thermal conductivity as a function of spherical 120 μm MgO content in a range of 50–75 vol%. A high filler loading above 50 vol% was indispensable to reaching above 1 W/mK. At a low concentration, the fillers were independently dispersed in a matrix and hardly contacted with each other. The thermal conductivity gradually increased as MgO content increased and reached the highest value of 1.67 W/mK at 60 vol%. The improved thermal conductivity was associated with the enhanced interconnectivity between the MgO particles. With an increase above 65 vol%, thermal conductivity decreased below 1.4 W/mK at which uniform dispersion is difficult to achieve because of high viscosity and agglomeration. Composition-dependent adhesion strength behaves similarly. The adhesives exhibited the highest value of 12.0 kgf/cm$^2$ at 60 vol% and 12.2 kgf/cm$^2$ at 65 vol% and then decreased thereafter with MgO loading (Figure 3b). It is commonly recognized that the agglomeration of inorganic particles can result in low adhesion strength. From the thermal conductivity and adhesion strength, the optimal MgO content for fluorosilicone adhesives is determined to be 60 vol%.

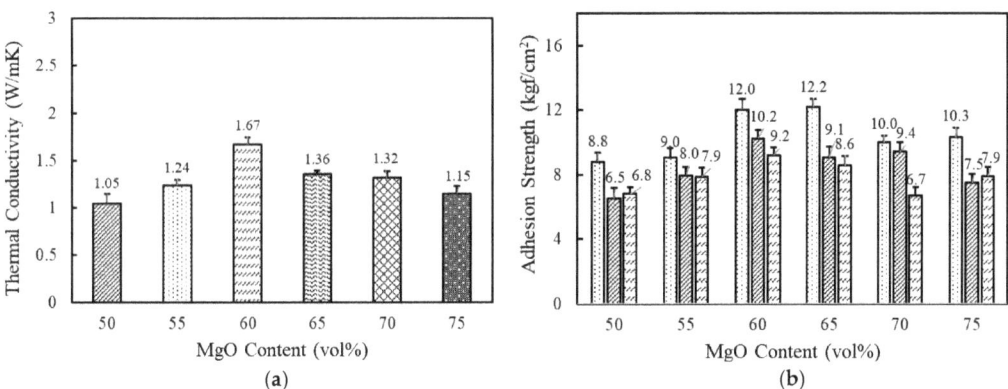

**Figure 3.** The thermal conductivity and adhesion strength of fluorosilicone adhesives as a function of MgO content: (**a**) thermal conductivity of fluorosilicone adhesives; (**b**) adhesion strength of fluorosilicone adhesives measured at 23 °C, 180 °C, and 250 °C.

*3.3. Thermal and Mechanical Properties of Bimodal Adhesives*

Hybrid adhesives containing more than two different fillers are often used to achieve high thermal conductivity while maintaining good processability. The combination of two or more fillers can help to form thermally conductive bridges at lower loadings by maximizing the packing density [33–35]. We selected spherical 120 μm MgO as a first component considering its high thermal conductivity, and the second component was added at a ratio of 45 vol% and 15 vol%. As shown in Figure 4a, adhesives consisting of 120 μm and 90 μm MgO, i.e., Bi-MgO-1, exhibited the highest thermal conductivity value of 1.82 W/mK, which was about 0.15 W/mK and 0.45 W/mK higher than those filled with the respective MgOs. Other bimodal adhesives containing 50 μm, 6 μm, and 0.6 μm MgOs as a second component also show higher thermal conductivity values compared to unimodal adhesives. Regarding the adhesion property, they revealed a similar strength over the whole temperature range measured regardless of MgO compositions (Figure 4b). The enlarged contact surface between fluorosilicone and MgO when using smaller second fillers may be responsible for the enhanced adhesion strength of unimodal 120 μm MgO.

Meanwhile, the adhesion strength of Bi-MgO-4 is slightly lower than that of unimodal 0.6 µm MgO.

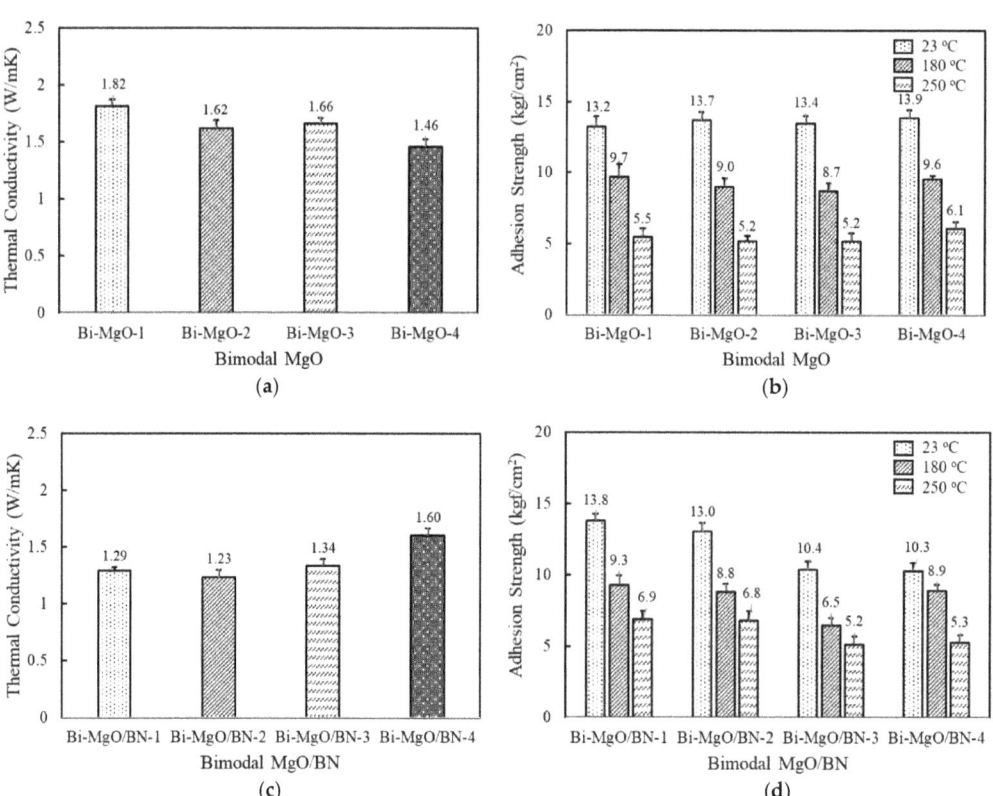

**Figure 4.** Thermal conductivity and adhesion strength of fluorosilicone adhesives containing two-component MgO and BN fillers: (**a**) thermal conductivity of bimodal MgO adhesives; (**b**) adhesion strength of bimodal MgO adhesives measured at 23 °C, 180 °C, and 250 °C; (**c**) thermal conductivity of bimodal MgO/BN adhesives; (**d**) adhesion strength of bimodal MgO/BN adhesives measured at 23 °C, 180 °C, and 250 °C. The filler concentration was kept at 60 vol%.

Figure 4c illustrates the thermal conductivity of bimodal MgO/BN adhesives. The second component BNs are combined with spherical 120 µm MgO to achieve a positive synergistic effect. The thermal conductivity of bimodal MgO/BN was increased above 1.2 W/mK because of the addition of a large amount of spherical MgOs (Figure 4c). The thermal conductivity of the bimodal MgO/BN was different from that of the single BN-filled adhesives. Bi-MgO/BN-4 containing 280 µm BN flakes exhibited the highest thermal conductivity value of 1.6 W/mK, followed by Bi-MgO/BN-3. In a single-component system, the aggregated and flake BNs showed low thermal conductivity below 1 W/mK. When a small amount of BNs (15 vol%) was added, they were uniformly dispersed without inducing phase separation and randomly orientated to build a three-dimensional conducting path along the thickness direction. Both filler size and shape determine the thermal conductivity of Bi-MgO/BN-3 and Bi-MgO/BN-4. The thermal conductivity of Bi-MgO-1 is not much different from that of Bi-MgO-2. The thermal conductivity results of bimodal MgOs and MgO/BN adhesives clearly indicated that the mixed use of large fillers with different shapes was effective in improving the filling density and building thermally conductive networks. As shown in Figure 4d, the adhesion strength of the bimodal MgO/BN showed an opposite behavior, exhibiting a high strength above 13 kgf/cm$^2$ for Bi-MgO/BN-1 and Bi-MgO/BN-2, while about 10 kgf/cm$^2$ for Bi-MgO/BN-3 and Bi-MgO/BN-4. The contact area between fluorosilicone resins and BNs mainly affected the adhesion strength of bimodal adhesives.

### 3.4. Thermal and Mechanical Properties of Trimodal Adhesives

The thermal conductivity of the trimodal adhesives filled with three component MgOs are displayed in Figure 5a. The composition of the bimodal MgO that exhibited the highest thermal conductivity, i.e., a combination of 120 μm and 90 μm, was partially substituted by the inclusion of the third component at 6.7 vol%. The thermal conductivity of trimodal MgOs was found to be in the range of 1.4~1.67 W/mK depending on the third component. Tri-MgO-2 containing amorphous 6 μm exhibited the highest thermal conductivity of 1.67 W/mK, while Tri-MgO-1 filled with spherical 50 μm MgO showed a thermal conductivity of 1.40 W/mK. Tri-MgO-1 should have high thermal conductivity considering its size but exhibited the lowest value. It is inferred that the addition of a third filler with a different shape is more effective to improve the packing efficiency. The thermal conductivity of trimodal MgO was slightly lower than Bi-MgO-1 because the enhanced thermal conductivity using smaller particles may be compensated by increased phonon scattering. When BN was used as the third component instead of MgO, no noticeable difference in thermal conductivity was observed with a value near 1.5 W/mK (Figure 5b). Since the amount of the third component was 6.7 vol%, the thermal conductivity may be dominated by a majority of spherical MgOs. In the case of trimodal adhesives containing two-component BNs along with 120 μm spherical MgO, the combined use of aggregated and flake-shaped BNs, i.e., Tri-MgO/BN2-3, was profitable to achieve high thermal conductivity, similarly to bimodal adhesives (Figure 5c).

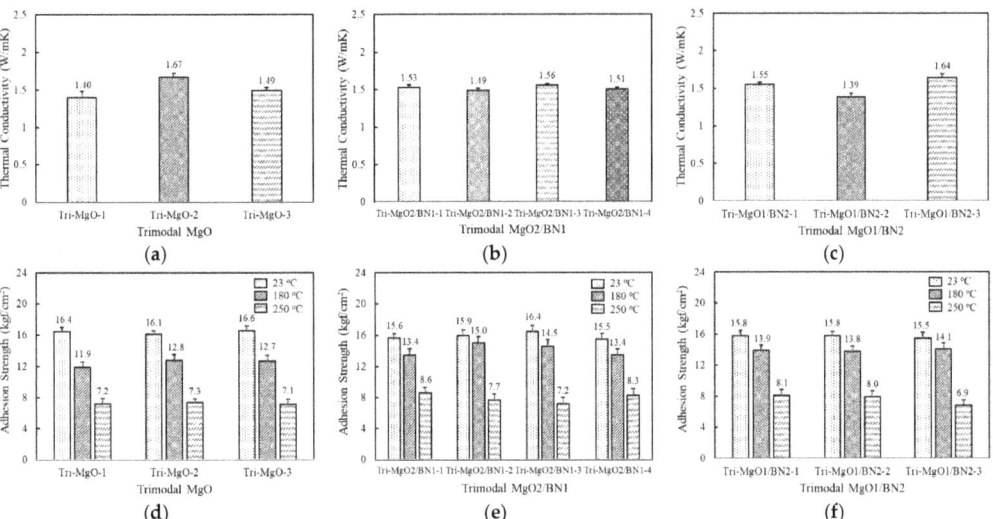

**Figure 5.** The thermal conductivity and adhesion strength of fluorosilicone adhesives containing three-component MgO and BN fillers: (**a**) thermal conductivity of trimodal MgO; (**b**) thermal conductivity of trimodal MgO2/BN1; (**c**) thermal conductivity of trimodal MgO1/BN2; (**d**) adhesion strength of trimodal MgO at 23 °C, 180 °C, 250 °C; (**e**) adhesion strength of trimodal MgO2/BN1 at 23 °C, 180 °C, 250 °C; (**f**) adhesion strength of trimodal MgO1/BN2 at 23 °C, 180 °C, nad250 °C.

The adhesion strength of trimodal adhesives is illustrated in Figure 5d–f. Although the adhesives were composed of various combinations of MgOs and BNs, the adhesion strength at 23 °C was quite similar, near 16 kgf/cm$^2$. Compared to unimodal and bimodal adhesives, a higher strength value was obtained because the third filler may enlarge the interaction area per unit gram. The adhesion strength at 180 °C fell in the range of 12.7~15.0 kgf/cm$^2$ depending on composition, which was comparable to those of unimodal and bimodal adhesives at 23 °C. At 250 °C, the strength value was maintained above 7.0 kgf/cm$^2$, except for Tri-MgO1/BN2-3.

Figure 6 shows the cross-sectional SEM images of various bimodal and trimodal adhesives. The adhesives filled with various MgOs, i.e., Figure 6a,b,d, exhibited only a thick layer of fluorosilicone resin on the fracture surface, which in turn interrupted the identification of the majority MgO components. It is inferred that the trimethoxy silane moiety presented on the surface of spherical MgO may offer the proper compatibility and good level of dispersion within fluorosilicone resins. When BNs were added, fillers with various shapes such as plate and agglomerates could be clearly discerned in bimodal and trimodal adhesives, as illustrated in Figure 6c,e,f. The shape is more apparent at the enlarged scale. Although the surface of BNs was not chemically modified, the agglomerates of fillers were not observed over the entire area. Therefore, mechanical mixing using a three-roll mill was effective for the uniform dispersion of micro-sized MgOs and BNs.

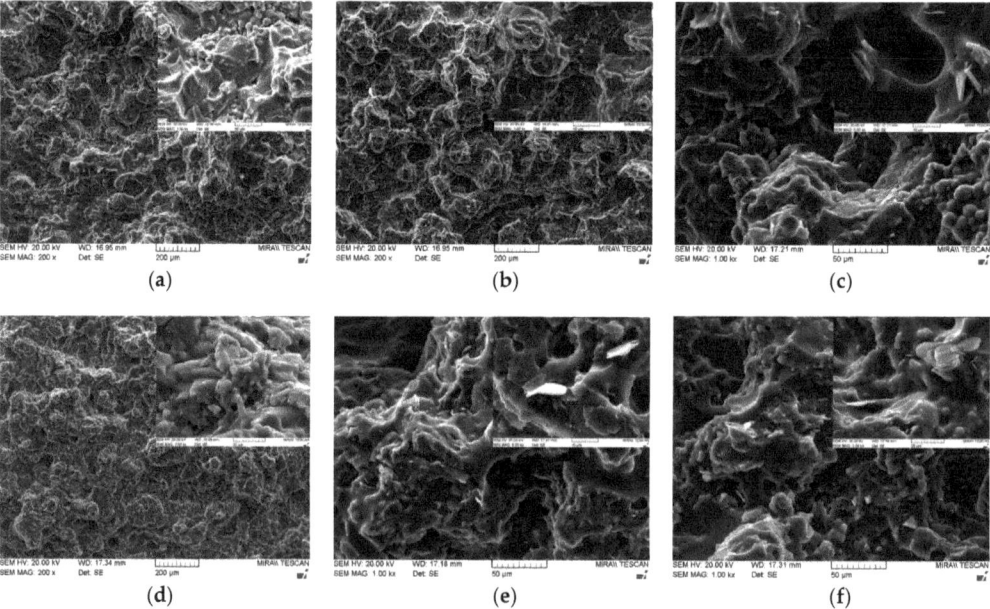

**Figure 6.** SEM images of bimodal and trimodal adhesives consisting of combined MgO and BN: (**a**) Bi-MgO-1 (120 μm MgO/90 μm MgO); (**b**) Bi-MgO-4 (120 μm MgO/0.6 μm MgO); (**c**) Bi-MgO/BN-4 (120 μm MgO/280 μm BN; (**d**) Tri-MgO-2 (120 μm MgO/90 μm MgO/6 μm MgO); (**e**) Tri-MgO2/BN1-3 (120 μm MgO/90 μm MgO/12 μm BN); (**f**) Tri-MgO1/BN2-3 (120 μm MgO/12 μm BN/280 μmBN).

The adhesion strength of unimodal, bimodal, and trimodal adhesives with relatively high thermal conductivity was compared before and after aging at 85 °C and 85% RH for 168 h. At the same volume fraction, the adhesion strength of MgO-1 containing only 120 μm MgO gradually increased when the smaller second and third MgOs were added, i.e., 13.2 kgf/cm$^2$ for Bi-MgO-1 and 16.1 kgf/cm$^2$ for Tri-MgO-2 (Figure 7a). The BN-filled trimodal adhesives such as Tri-MgO2/BN1-3 and Tri-MgO1/BN2-3 also showed a high adhesion strength above 15 kgf/cm². Smaller fillers may occupy the microscopic gaps between large MgO particles and consequently increase the number of joint points. The adhesion strength obtained after humid and thermal aging is illustrated in Figure 7b. Additional cross-linking reactions may take place for fluorosilicone during a course of aging, leading to enhanced adhesion strength for bimodal and trimodal adhesives. Tri-MgO2/BN1-3 exhibited the highest strength of 22.4 kgf/cm² at 23 °C, 15.6 kgf/cm² at 180 °C, and 13.3 kgf/cm² at 250 °C. It should be noted that the sequence of adhesion strength of unimodal, bimodal, and trimodal adhesives before and after aging is almost unchanged. A higher adhesion strength was still maintained at 180 °C and 250 °C.

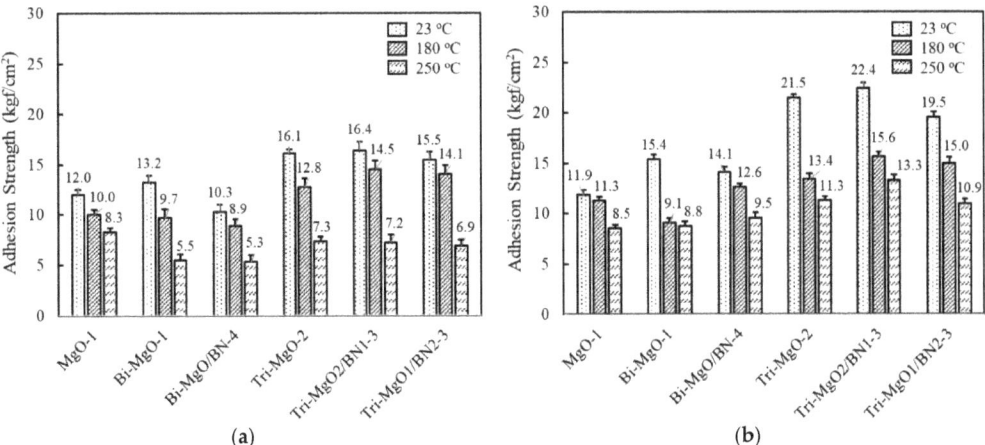

**Figure 7.** Comparison of the adhesion strength of unimodal, bimodal, and trimodal adhesives before and after aging at 85 °C/85% RH for 168 h: (**a**) Adhesion strength obtained before aging; (**b**) adhesion strength obtained after aging.

## 4. Conclusions

The effects of the filler size, shape, and their combination on the thermal conductivity and adhesion strength of fluorosilicone was demonstrated using MgO and BN fillers. The use of two or three different fillers can lead to high thermal conductivity and adhesion strength. Bimodal adhesives filled with spherical 120 μm MgO and 90 μm MgO exhibited the highest thermal conductivity due to the synergistic effect of high packing density and low interfacial resistance. In trimodal adhesives, the enhanced packing density achieved by using different shapes is important to determine the thermal conductivity. Aggregated and flake shaped BNs alone would be inadequate to achieve high thermal conductivity, but are effective as secondary fillers. Trimodal adhesives exhibited high adhesion strength as compared to unimodal and bimodal adhesives even after thermal and humid aging. Fluorosilicone adhesives have promising applications in electronics considering their chemical and thermal resistance, although their adhesion strength is relatively low.

**Author Contributions:** Conceptualization, K.-S.S. and N.K.; Methodology, S.-Y.K. and M.-K.O.; Formal Analysis, S.-Y.K. and K.-S.S.; Investigation, M.-K.O. and N.K.; Writing Review & Editing, N.K.; Visualization, K.-S.S.; Supervision, N.K. All authors have read and agreed to the published version of the manuscript.

**Funding:** This research was funded by the Ministry of Trade, Industry, and Energy grand number 20014374.

**Institutional Review Board Statement:** Not applicable.

**Informed Consent Statement:** Not applicable.

**Data Availability Statement:** Not applicable.

**Acknowledgments:** This work was supported by the Development of Nano Convergence Innovative Products Project of the Ministry of Trade, Industry, and Energy (grant number 20014374).

**Conflicts of Interest:** The authors declare no conflict of interest.

## References

1. Yim, M.J.; Paik, K.W. Review of Electrically Conductive Adhesive Technologies for Electronic Packaging. *Electron. Mater. Lett.* **2006**, *2*, 183–194.
2. Huang, X.; Jiang, P.; Tanaka, T. A Review of Dielectric Polymer Composites with High Thermal Conductivity. *IEEE Electr. Insul. Mag.* **2011**, *27*, 8–16. [CrossRef]
3. Procter, P.; Solc, J. Improved thermal conductivity in microelectronic encapsulants. *IEEE Trans. Compon. Hybrids Manuf. Technol.* **1991**, *14*, 708–713. [CrossRef]
4. Kim, Y.H.; Lim, Y.W.; Lee, D.W.; Kim, Y.H.; Bae, B.S. A highly adhesive siloxane LED encapsulant optimized for high themal stability and optical efficiency. *J. Mater. Chem. C* **2016**, *4*, 10791–10796.
5. Wägli, P.; Homsy, A.; de Rooij, N.F. Norland optical adhesive (NOA81) microchannels with adjustable wetting behavior and high chemical resistance against a range of mid-infrared-transparent organic solvents. *Sens. Actuators B Chem.* **2011**, *156*, 994–1001.
6. Marques, E.A.S.; da Silva, L.F.M.; Banea, M.D.; Carbas, R.J.C. Adhesive Joints for Low- and High-Temperature Use: An Overview. *J. Adhes.* **2015**, *91*, 556–585. [CrossRef]
7. Jojibabu, P.; Zhang, Y.X.; Gangadhara Prusty, B. A review of research advances in epoxy-based nanocomposites as adhesive materials. *Int. J. Adhes. Adhes.* **2020**, *96*, 102454. [CrossRef]
8. Ahmadi, Z. Nanostructured epoxy adhesives: A review. *Prog. Org. Coat.* **2019**, *135*, 449–453. [CrossRef]
9. Jin, F.L.; Li, X.; Park, S.J. Synthesis and application of epoxy resins: A review. *J. Ind. Eng. Chem.* **2015**, *29*, 1–11. [CrossRef]
10. Ramos, J.A.; Pagani, N.; Riccardi, C.C.; Borrajo, J.; Goyanes, S.N.; Mondragon, I. Cure kinetics and shrinkage model for epoxy-amine systems. *Polymer* **2005**, *46*, 3323–3328. [CrossRef]
11. Hsieh, T.H.; Kinloch, A.J.; Masania, K.; Lee, J.S.; Taylor, A.C.; Sprenger, S. The toughness of epoxy polymers and fibre composites modified with rubber microparticles and silica nanoparticles. *J. Mater. Sci.* **2010**, *45*, 1193–1210. [CrossRef]
12. Kumar, S.A.; Balakrishnan, T.; Alagar, M.; Denchev, Z. Development and characterization of silicone/phosphorus modified epoxy materials and their application as anticorrosion and antifouling coatings. *Prog. Org. Coat.* **2006**, *55*, 207–217. [CrossRef]
13. Ben Saleh, A.B.; Mohd Ishak, Z.A.; Hashim, A.S.; Kamil, W.A.; Ishiaku, U.S. Synthesis and characterization of liquid natural rubber as impact modifier for epoxy resin. *Phys. Procedia* **2014**, *55*, 129–137. [CrossRef]
14. Liu, L.; Kuffel, K.; Scott, D.K.; Constantinescu, G.; Chung, H.J.; Rieger, J. Silicone-based adhesives for long-term skin application: Cleaning protocols and their effect on peel strength. *Biomed. Phys. Eng. Express* **2018**, *4*, 015004. [CrossRef]
15. Staudt, Y.; Odenbreit, C.; Schneider, J. Failure behaviour of silicone adhesive in bonded connections with simple geometry. *Int. J. Adhes. Adhes.* **2018**, *82*, 126–138. [CrossRef]
16. Li, Z.; Le, T.; Wu, Z.; Yao, Y.; Li, L.; Tentzeris, M.; Moon, K.S.; Wong, C.P. Rational Design of a Printable, Highly Conductive Silicone-based Electrically Conductive Adhesive for Stretchable Radio-Frequency Antennas. *Adv. Funct. Mater.* **2015**, *25*, 464–470. [CrossRef]
17. Cornelius, D.J.; Monroe, C.M. The Unique Properties of Silicone and Fluorosilicone Elastomers. *Polym. Eng. Sci.* **1985**, *25*, 467–473. [CrossRef]
18. Zheng, X.; Pang, A.; Wang, Y.; Wang, W.; Bai, Y. Fabrication of UV-curable fluorosilicone coatings with impressive hydrphobicity and solvent resistance. *Prog. Org. Coat.* **2020**, *144*, 105633. [CrossRef]
19. Bernstein, R.; Gillen, K.T. Predicting the lifetime of fluorosilicone o-rings. *Polym. Degrad. Stab.* **2009**, *94*, 2107–2113. [CrossRef]
20. Hwang, Y.J.; Kim, J.M.; Kim, L.S.; Jang, J.Y.; Kim, M.S.; Jeong, S.Y.; Cho, J.Y.; Yi, G.R.; Choi, Y.S.; Lee, G.H. Epoxy-based thermally conductive adhesives with effective alumina and boron nitride for superconducting magnet. *Compos. Sci. Technol.* **2020**, *200*, 108456. [CrossRef]
21. Cui, H.W.; Li, D.S.; Fan, Q. Using Nano Hexagonal Boron Nitride Particles and Nano Cubic Silicon Carbide Particles to Improve the Thermal Conductivity of Electrically Conductive Adhesives. *Electron. Mater. Lett.* **2013**, *9*, 1–5. [CrossRef]
22. Mai, V.D.; Lee, D.I.; Park, J.H.; Lee, D.S. Rheological Properties and Thermal Conductivity of Epoxy Resins Filled with a Mixture of Alumina and Boron Nitride. *Polymers* **2019**, *11*, 597. [CrossRef]
23. Yetgin, H.; Veziroglu, S.; Aktas, O.C.; Yalçinkaya, T. Enhancing thermal conductivity of epoxy with a binary filler system of h-BN platelets and $Al_2O_3$ nanoparticles. *Int. J. Adhes. Adhes.* **2020**, *98*, 102540. [CrossRef]
24. Bian, W.; Yao, T.; Chen, M.; Zhang, C.; Shao, T.; Yang, Y. The synergistic effects of the micro-BN and nano-$Al_2O_3$ in micro-nano composites on enhancing the thermal conductivity for insulating epoxy resin. *Compos. Sci. Technol.* **2018**, *168*, 420–428. [CrossRef]
25. Choi, S.; Kim, J. Thermal conductivity of epoxy composites with a binary-particle system of aluminum oxide and aluminum nitride fillers. *Compos. Part B* **2013**, *51*, 140–147. [CrossRef]
26. Hong, J.P.; Yoon, S.W.; Hwang, T.S.; Oh, J.S.; Hong, S.C.; Lee, Y.K.; Nam, J.D. High thermal conductivity epoxy composites with bimodal distribution of aluminum nitride and boron nitride fillers. *Thermochim. Acta* **2012**, *537*, 70–75. [CrossRef]

27. Kim, K.H.; Kim, M.J.; Kim, J.H. Thermal and mechanical properties of epoxy composites with a binary particle filler system consisting of aggregated and whisker type boron nitride particles. *Compos. Sci. Technol.* **2014**, *103*, 72–77. [CrossRef]
28. Wereszczak, A.A.; Morrissey, T.G.; Volante, C.N. Thermally Conductive MgO-Filled Epoxy Molding Compounds. *IEEE Trans. Compon. Packag. Manuf. Technol.* **2013**, *3*, 1994–2005. [CrossRef]
29. Mohammad, W.A.; Kim, J.S.; Muddassir, A.M.; Muhammad, Y.K.; Muhammad, M.B. Surface modification of magnesium oxide/epoxy composites with significantly improved mechanical and thermal properties. *J. Mater. Sci. Mater. Electron.* **2021**, *32*, 15307–15316.
30. Donnay, M.; Tzavalas, S.; Logakis, E. Boron nitride filled epoxy with improved thermal conductivity and dielectric breakdown strength. *Compos. Sci. Technol.* **2015**, *110*, 152–158. [CrossRef]
31. Xie, B.H.; Huang, X.; Zhang, G.J. High thermal conductive polyvinyl alcohol composites with hexagonal boron nitride microplatelets as fillers. *Compos. Sci. Technol.* **2013**, *85*, 98–103. [CrossRef]
32. Lin, Z.; Liu, Y.; Raghavan, S.; Moon, K.S.; Sitaraman, S.K.; Wong, C.P. Magnetic alignment of hexagonal boron nitride platelets in polymer matrix: Toward high performance anisotropic polymer composites for electronic encapsulation. *ACS Appl. Mater. Interfaces* **2013**, *5*, 7633–7640. [CrossRef] [PubMed]
33. Kochetov, R.; Korobko, A.V.; Andritsch, T.; Morshuis, P.H.F.; Picken, S.J. Modelling of the thermal conductivity in polymer nanocomposites and the impact of the interface between filler and matrix. *J. Phys. D Appl. Phys.* **2011**, *44*, 395401. [CrossRef]
34. Liu, C.; Chen, M.; Zhou, D.; Wu, D.; Yu, W. Effect of Filler Shape on the Thermal Conductivity of Thermal Functional Composites. *J. Nanomater.* **2017**, *2017*, 1–15. [CrossRef]
35. Zhao, Y.; Zhai, Z.; Drummer, D. Thermal Conductivity of Aluminosilicate- and Aluminum Oxide-Filled Thermosets for Injection Molding: Effect of Filler Content, Filler Size and Filler Geometry. *Polymers* **2018**, *10*, 457. [CrossRef]

*Article*

# Near-Infrared Light-Responsive Shape Memory Polymer Fabricated from Reactive Melt Blending of Semicrystalline Maleated Polyolefin Elastomer and Polyaniline

Min-Su Heo [1], Tae-Hoon Kim [1,2], Young-Wook Chang [1,2,*] and Keon Soo Jang [3]

[1] Department of Materials & Chemical Engineering, Hanyang University, Ansan 15588, Gyeonggi-do, Korea; su920122@hanyang.ac.kr (M.-S.H.); taehoonkim@hanyang.ac.kr (T.-H.K.)
[2] BK21 FOUR ERICA-ACE Center, Hanyang University, Ansan 15588, Gyeonggi-do, Korea
[3] Department of Polymer Engineering, School of Chemical and Materials Engineering, The University of Suwon, Hwaseong 18323, Gyeonggi-do, Korea; ksjang@suwon.ac.kr
* Correspondence: ywchang@hanyang.ac.kr; Tel.: +82-31-400-5277

**Abstract:** A shape memory polymer was prepared by melt mixing a semicrystalline maleated polyolefin elastomer (mPOE) with a small amount of polyaniline (PANI) (up to 15 wt.%) in an internal mixer. Transmission electron microscopy (TEM), FTIR analysis, DMA, DSC, melt rheological analysis, and a tensile test were performed to characterize the structure and properties of the mPOE/PANI blends. The results revealed that the blends form a physically crosslinked network via the grafting of PANI onto the mPOE chains, and the PANI dispersed at the nanometer scale in the POE matrix served as a photo-thermal agent and provided increased crosslinking points. These structural features enabled the blends to exhibit a shape memory effect upon near-infrared (NIR) light irradiation. With increasing PANI content, the shape recovery rate of the blend under NIR stimulation was improved and reached 96% at 15 wt.% of PANI.

**Keywords:** shape memory polymer; NIR light responsive; semicrystalline maleated polyolefin elastomer; polyaniline; melt blending

## 1. Introduction

A shape memory polymer (SMP) is a smart material that can memorize its original shape and recover to its original shape from temporarily fixed shapes, and responds to various types of external stimuli, such as heat, light, water, and electric or magnetic fields. Compared to metallic shape memory alloys, SMPs have several advantages, including low density, good processability, high deformability, biocompatibility, as well as easy control of transition temperature, which make these materials have various applications [1–5]. More extensive applications of SMPs were reported very recently, which include fabrications of smart surfaces with adjustable wetting properties [6], 3D structures with various complicated shapes [7], 4D-printed medical devices [8], and deployable solar arrays [9].

Among the various types of SMPs, light-responsive SMPs have gained interest due to several advantages they possess over conventional, thermally activated SMPs, such as remote activation ability and spatial controllability [10]. The most frequently employed approach to fabricate light-responsive SMPs is the incorporation of small amounts of photothermal reagents, such as gold [11], silver [12], titanium nitride [13], polydopamine [14], porphyrin [15], or carbon nanoparticles, such as CNT and graphene [16–19], into the shape memory polymer matrix. These additives act as photothermal heaters when the composites are irradiated with near infrared or sunlight, which enables the SMP nanocomposites to be indirectly heated above their transition temperature. For these composite systems, however, proper surface modifications of the fillers are required to enable them to be finely dispersed in matrix polymers.

Conjugated polymers, such as polyaniline, polypyrrole, and polythiophene, show strong absorption of near-infrared energy, and transform it into thermal energy with high efficiency, which enables them to be used in photothermal therapy [20–23]. Only a few studies have been reported on light-responsive SMPs using these conjugated polymers as a photothermal agent [24–26]. Bai et al. reported an NIR-induced shape memory hydrogel by the incorporation of PANI nanofibers into polyvinyl alcohol (PVOH) by in situ polymerization of aniline in PVOH, where the PANI nanofiber served as a photothermal agent and provided increased crosslinking points [24]. Coating the PANI nanofiber onto a shape memory epoxy was also reported to exhibit NIR-responsive shape memory behavior [25]. Polypyrrole was incorporated into a shape memory polyurethane to prepare an NIR-induced shape memory elastomer [26]. Despite their excellent light-responsive behaviors, solution processing was needed to fabricate these SMPs, which was complicated and time consuming. It would be more practically useful if the photo-responsive SMPs based on conjugated polymers could be fabricated by a melt blending method.

PANI, which has a sulfonic acid group on its aromatic ring, is of particular interest, since it has a self-doped structure without the addition of any external dopants and exhibits strong absorption in the near-infrared (NIR) region [27–29], which makes it a promising photothermal agent. However, its use as a photothermal agent for the fabrication of light-responsive shape memory polymers has not been explored, especially by using a melt blending method. In this study, we observed that PANI obtained from oxidative chemical polymerization of aniline, possessing a sulfonic acid, can be covalently bonded onto a semicrystalline polyolefin elastomer, possessing a maleic anhydride group (mPOE), during melt processing, and is dispersed in the elastomer matrix at the nanometer scale. Further, the mPOE/PANI blends formed a physically crosslinked structure and exhibited photothermal behavior under NIR irradiation, which enabled these blends to have good light-responsive shape memory effects. Along with the examinations of photo-responsive shape memory effects, the thermo-mechanical and rheological properties of the blends were also investigated.

## 2. Materials and Methods

*2.1. Materials and Sample Preparation*

Semicrystalline maleated polyolefin elastomer (mPOE, maleic anhydride content 2 wt.%) was procured from DuPont (Fusabond N416, Wilmington, DE, USA). Aniline-3-sulfonic acid (ANISA, >99.0%) and ammonium persulfate (APS, 95%) were purchased from Sigma-Aldrich (St. Louis, MO, USA). Pyridine and acetone were procured from Junsei Chemical Co., Ltd. (Tokyo, Japan).

Polyaniline (PANI) was synthesized by oxidative chemical polymerization using ANISA as a monomer and APS as an oxidant as reported in the literature [27]. Briefly, 35.0 g (0.2 mol) of ANISA was dissolved in 100 mL of 3.0 mol/L pyridine aqueous solution at 4 °C. Further, 45.6 g (0.2 mol) of APS was dissolved in 180 mL of water. The APS solution was slowly added to the ANISA solution at a temperature below 4 °C for 1 h, and then the mixture was stirred for 15 h. The precipitate was collected and washed in acetone, and then dried under vacuum for 24 h at 30 °C.

mPOE/PANI blends with 3, 5, 10 and 15 wt.% PANI were prepared by melt blending in a Haake mixer (Haake Polylab Rheomix 600, Germany) for 10 min at 180 °C with a rotor speed of 60 rpm. The resulting mixtures were then molded as sheets by using an electrically heated hydraulic press at 180 °C for 10 min.

*2.2. Characterization*

Phase morphology of the blends was examined using transmission electron microscopy (TEM), with a JEOL 200CX TEM with an acceleration voltage of 200 kV. Ultra-thin sections were cut from the compression-molded specimens with a thickness of 100 nm using a Reichert ultracut cryo-microtome.

FTIR analysis was carried out using Bucker ALPHA spectrometer (Nicolet IS10, Thermo scientific, Waltham, MA, USA), equipped with an attenuated total reflectance (ATR) accessory, to characterize the PANI, and evaluate the possible interactions between the mPOE and PANI as well. The sample was placed on the ATR sample holder. The number of scans for each spectra was 16. The FTIR spectra were obtained in the range of 2000–700 cm$^{-1}$ under nitrogen atmosphere. The empty sample chamber and ATR stage were used to obtain the background spectra.

Dynamic mechanical tests were carried out using a dynamic mechanical analyzer (TA Instrument, model DMA-Q800, Waltham, MA, USA). Samples were subjected to a cyclic tensile strain with an amplitude of 0.2% at a frequency of 1 Hz. The temperature was increased at a heating rate of 10 °C/min over the range from −100 to 150 °C.

Thermal properties of the samples were examined by using a differential scanning calorimeter (TA instruments, DSC Q20 equipped with RSC90 as a refrigerated cooling system). Samples were first heated from 30 to 200 °C at a rate of 10 °C/min under a nitrogen atmosphere and were kept for 5 min at this temperature to remove prior thermal history. The samples were then cooled down to −50 °C at a cooling rate of 10 °C/min (cooling scan), and were reheated to 200 °C at the same heating rate (second heating scan).

Tensile properties of the samples were measured by using a universal testing machine (UTM, AGS-500NX, Shimadzu, Japan) with a crosshead speed of 50 mm/min. The test was repeated at least five times at room temperature for each sample.

Melt rheological measurements were conducted on an MCR 102 rheometer (Anton paar, Graz, Austria). Dynamic oscillatory shear measurements were performed at 180 °C using parallel plate geometry with a plate diameter of 25 mm and 1–2-mm-thick samples. The dynamic frequency sweep test was carried out between 0.1 and 100 rad/s.

Photothermal effects of the blend were examined by measuring the surface temperature of the NIR-exposed sample using a digital thermometer during the NIR irradiation of the sample with 805 nm NIR at power density of 1.0 W/cm$^2$ positioned at 30 cm from the sample. An infrared (IR) lamp with a red filter (Philips, Model Infraphil PAR38E, Berlin, Germany) was used as a light source.

Shape memory behavior of the samples was evaluated using dog-bone-shaped specimens with a thickness of about 1 mm. Temporarily fixed sample was obtained by uniaxial deformation of the specimen to 50% at 70 °C (which is just above $T_m$ of the sample) followed by cooling the deformed shape to 0 °C. Then, the temporarily deformed samples were exposed to NIR irradiation to analyze the shape recovery. The shape fixing ratio ($R_f$) and shape recovery ratio ($R_r$) of samples were determined by the following equations:

$$R_f\ (\%) = [(L_s - L_i)/(L_u - L_i)] \times 100 \tag{1}$$

$$R_r\ (\%) = [(L_s - L_r)/(L_s - L_i)] \times 100 \tag{2}$$

where $L_i$ denotes the initial gauge length of the sample, $L_u$ denotes the stretched length under load (1.5 × $L_i$ in this experiment), $L_s$ denotes the stretched length without load, and $L_r$ denotes the recovered length upon NIR exposure, respectively.

## 3. Results

### 3.1. Phase Morphology

The phase morphology of the mPOE/PANI blends examined by TEM, and the TEM micrographs for the blends with various PANI contents are shown in Figure 1. A phase-separated morphology could be observed for all the samples in which the number of dispersed domains and their average diameter increased with increasing PANI contents in the blend, from about 100 nm for the blend with 3 wt.% of PANI to ca. 200 nm for the blend with 15 wt.% of PANI.

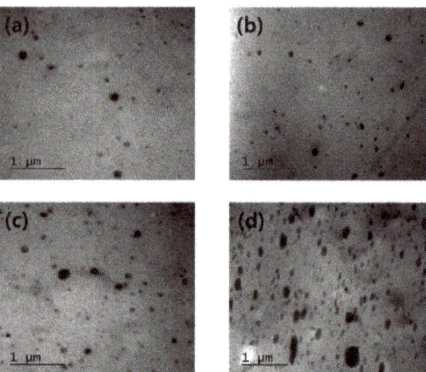

**Figure 1.** FE-TEM images of mPOE/PANI blends with (**a**) 3, (**b**) 5, (**c**) 10, and (**d**) 15 wt.% of PANI content (magnification ×50.0 k).

*3.2. FT-IR*

FTIR characterization was performed to examine any possible interactions between mPOE and PANI in the blend. Figure 2 shows the FTIR spectra of neat mPOE and the mPOE/PANI (90/10) blend. The FTIR spectra of neat mPOE show characteristic absorption peaks at 1790 cm$^{-1}$ and 1866 cm$^{-1}$, attributed to the stretching vibration of cyclic maleic anhydride. The band at 1712 cm$^{-1}$ was contributable to the stretching vibration of the carbonyl moiety attached to the cyclic maleic anhydride group [30]. In the FTIR spectra for the mPOE/PANI blend, a new absorption band, corresponding to 1652 cm$^{-1}$, appeared, attributed to the C=O stretching of amide bonds, while there was a decrease in peak intensity, corresponding to maleic anhydride. This indicates that a chemical reaction occurred between the maleic anhydride of mPOE and the amine of PANI during the melt mixing process at high temperature. Such a reaction between PANI and the maleic anhydride group of other maleated polymers was also observed in blends of PANI with maleated PBT and maleated PS, in which PANI was grafted onto the polymer chains via the reaction [31,32]. It is thought that the nanoscaled phase-separated morphology achieved in the mPOE/PANI blends studied here was led by the reaction between the component polymers during melt blending. The nanostructured polymer blends obtained by reactive blending have been reported in other polymer blend systems, such as nylon/polyolefin blends [33,34] and PLA/PBAT blends [35].

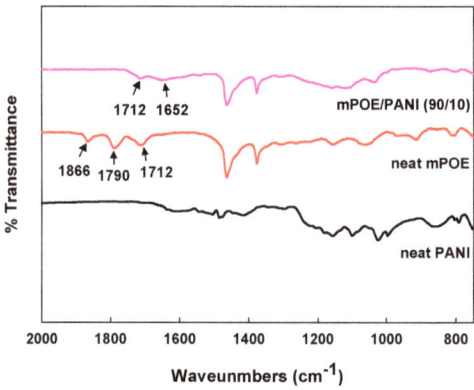

**Figure 2.** FT-IR spectra of neat PANI, neat mPOE and mPOE/PANI (90/10) blend.

## 3.3. Dynamic Mechanical Properties

Figure 3 shows the temperature dependency of storage modulus (E′) for neat mPOE and mPOE/PANI blends. The storage moduli for the blends are higher than those of the neat mPOE over the whole temperature range examined here, and the modulus increases with increasing PANI content in the blends, implying that the PANI nanoparticles acted as reinforcing fillers. It should also be noted that the storage modulus for neat mPOE decreases sharply when the temperature is higher than around 70 °C (which corresponds to $T_m$), whereas the modulus of mPOE/PANI blends keeps quite stable up to 200 °C. Such a persistent modulus indicates that the blends form a crosslinked structure, which was obtained via the chemical reaction between the mPOE and PANI in the blends, as was confirmed by FTIR analysis.

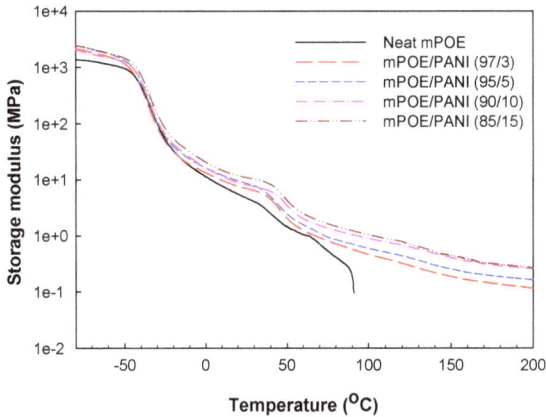

**Figure 3.** Variation in storage modulus (E′) with temperature for neat mPOE and mPOE/PANI blends.

## 3.4. Thermal Properties

DSC analysis was performed to observe the variation in melting temperature ($T_m$) and heat of fusion ($\Delta H_m$) with the composition of the blends, and those values and the associated degree of crystallinity ($\chi_c$) of the samples are listed in Table 1. The degree of crystallinity was calculated using $\Delta H_m$ per gram of mPOE obtained from DSC measurements and the $\Delta H_m$ corresponding to 100% crystalline low-density polyethylene (LDPE) (293 J/g) [30]. As illustrated in Table 1, the $T_m$ and $\chi_c$ of the mPOE phase of the blends were continuously decreased upon increasing the PANI content in the blends. This suggests that the ordered crystalline structure of the semicrystalline mPOE was disturbed in the mPOE/PANI blends due to the formation of a crosslinked structure, which caused a decrease in the molecular mobility of the POE chains, and, thus, the crystallization is restricted, as manifested by the decrease in $T_m$ and $\chi_c$ [36,37].

**Table 1.** Thermal characteristics of mPOE/PANI blends.

| Sample | $T_m$ (°C) | $\Delta H_m$ (J/g) | $\chi_c$ (%) [1] |
|---|---|---|---|
| Neat mPOE | 45.2 | 19.9 | 6.8 |
| mPOE/PANI (97/3) | 44.4 | 18.6 | 6.3 |
| mPOE/PANI (95/5) | 43.8 | 17.1 | 5.8 |
| mPOE/PANI (90/10) | 43.6 | 15.3 | 5.2 |
| mPOE/PANI (85/15) | 43.2 | 14.8 | 5.1 |

[1] $\chi_c = 100 \times (\Delta H_m / \Delta H_m°)/w$. $\Delta H_m$ and $\Delta H_m°$ are heats of melting for the sample and 100% crystalline LDPE (293 J/g), respectively. w: mass of the sample.

## 3.5. Tensile Properties

The stress–strain curves of neat mPOE and the mPOE/PANI blends are shown in Figure 4, and the results are summarized in Table 2. The tensile strength and modulus at a given strain of the blends are higher than those of neat mPOE, and the values increased with increasing PANI content in the blends. Neat mPOE exhibits typically elastomeric behavior, undergoing large deformation (about 1000% strain) with a low modulus. The blends exhibited similar deformation, along with a higher tensile modulus and strength, without serious loss of elongation at break, compared to neat mPOE. Such enhanced tensile properties are ascribed to the rigid nature of the PANI dispersed in the elastomer matrix, along with the strong interfacial adhesion between the two phases formed by covalent bonding, as discussed in the section above.

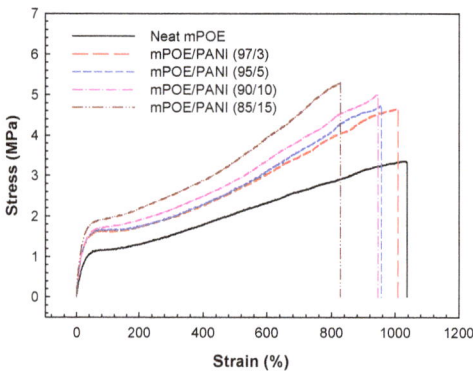

**Figure 4.** Stress–strain curves of mPOE/ PANI blends.

**Table 2.** Tensile properties of mPOE/PANI blends.

| Sample | 100% Tensile Modulus (MPa) | 300% Tensile Modulus (MPa) | Tensile Strength (MPa) | Elongation-at-Break (%) |
|---|---|---|---|---|
| Neat mPOE | 1.2 ± 0.1 | 1.5 ± 0.1 | 3.3 ± 0.2 | 1030 ± 70 |
| mPOE/PANI (97/3) | 1.6 ± 0.1 | 2.0 ± 0.2 | 4.6 ± 0.3 | 1000 ± 70 |
| mPOE/PANI (95/5) | 1.7 ± 0.1 | 2.0 ± 0.2 | 4.6 ± 0.3 | 950 ± 50 |
| mPOE/PANI (90/10) | 1.7 ± 0.1 | 2.2 ± 0.2 | 5.0 ± 0.3 | 940 ± 50 |
| mPOE/PANI (85/15) | 1.9 ± 0.1 | 2.5 ± 0.2 | 5.3 ± 0.3 | 820 ± 40 |

## 3.6. Melt Rheological Properties

It was observed that the blends are melt processable, even though they form a crosslinked structure, which suggests that the blends formed a physically crosslinked structure. In order to observe the effect of blend composition on melt rheological properties, melt viscosity was measured as a function of frequency by an oscillatory rheometer, and the results are presented in Figure 5. It can be observed that the melt viscosity of the mPOE/PANI blends showed a higher value compared to neat mPOE, especially at low frequencies, and it increased with increasing PANI content in the blends. Such enhanced viscosity of the mPOE/PANI blend indicated that the flow of POE chains in the molten state was restricted by the rigid PANI nanoparticles dispersed in the POE matrix. Additionally, all the blends showed typical shear-thinning behavior, in which viscosity decreases with increasing frequency, which is pronounced by increasing the PANI content in the blends. In general, the shear-thinning behavior is more pronounced in solid-like materials, such as the crosslinked polymer and polymer composites [38,39].

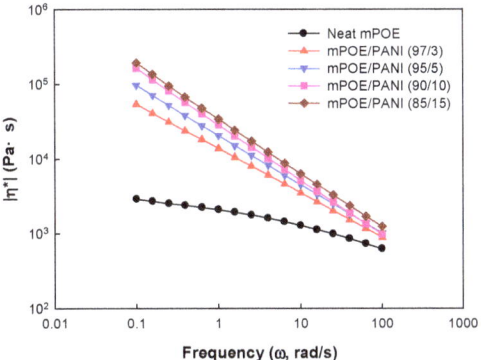

**Figure 5.** Variation in complex viscosity of mPOE/PANI blends with frequency.

## 3.7. Photothermal Effects

A photothermal effect occurs when the exposed light is absorbed by a substance and is converted into thermal energy, thereby increasing the substance temperature. The photothermal effect of the samples was observed by measuring the surface temperature under irradiation of NIR light (1.0 W/cm$^2$). Figure 6 shows the surface temperature of the pristine mPOE and mPOE/PANI blends as a function of NIR exposure time. The surface temperatures of the mPOE/PANI (85/15) blends increased from room temperature to 82 °C within 10 min of exposure, whereas the surface temperature of the neat mPOE only increased slightly. The results reveal that the PANI nano-domains dispersed in the POE matrix worked as photothermal nanoheaters, converting the imposed NIR to thermal energy.

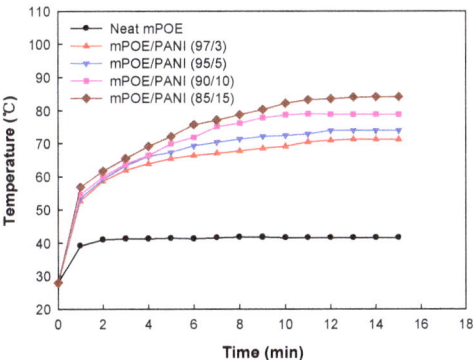

**Figure 6.** Variation in surface temperature with NIR irradiation time of mPOE/PANI blends.

## 3.8. NIR Light-Induced Shape Memory Effect

As discussed above, the mPOE/PANI blends formed a crosslinked structure via the chemical reaction between PANI and semicrystalline maleated POE, and the modulus in the plateau region, which is related with crosslink density, increased with increasing PANI content. Further, the mPOE/PANI blends exhibited higher photothermal behavior with increasing PANI content. These results indicate that PANI acted as a crosslinking agent for the semicrystalline elastomer, as well as a photothermal nanoheater. These structural features of the blends enable them to exhibit shape memory behavior, i.e., crosslinked points stabilize the permanent shape, and a reversible phase, which has a crystalline melting transition, serves as a switch. A tensile recovery test was used to evaluate the light-induced shape memory effects of the samples. Shape recovery was observed to occur when the temporarily fixed specimen was exposed to NIR light, and the results are demonstrated in Figure 7, and the shape fixity ($R_f$) and shape recovery ($R_r$), calculated according to Equations (1) and (2), respectively, are shown in Table 3. It can be observed that $R_r$ increased with increasing PANI content in the blends, and reached 96% for the blend with a PANI content of 15 wt.%. The improved shape recovery of the blends with a higher PANI content is ascribed to the increased crosslinked points, as well as the higher photothermal effects, as shown in the sections above. When the blend sample was deformed under a tensile load above its $T_m$, elastic energy was stored during this deformation. Photothermal heating led to the melting of the crystalline domains of the temporarily fixed sample, and the polymer chains recovered to its original shape by releasing the stored energy.

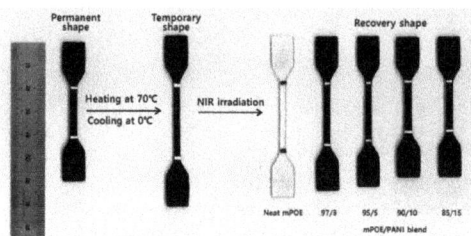

**Figure 7.** NIR light-induced shape recovery process of mPOE/ PANI blends.

**Table 3.** Shape memory effects of mPOE/PANI blends.

| Sample | $R_f$ (%) | $R_r$ (%) |
|---|---|---|
| Neat mPOE | 98.2 | 28.5 |
| mPOE/PANI (97/3) | 99.6 | 44.6 |
| mPOE/PANI (95/5) | 99.7 | 88.5 |
| mPOE/PANI (90/10) | 99.7 | 92.9 |
| mPOE/PANI (85/15) | 99.6 | 96.2 |

## 4. Conclusions

This study demonstrated that an NIR light-responsive shape memory elastomer can be fabricated by simple melt blending of a maleated semicrystalline polyolefin elastomer, with a small amount of polyaniline. Studies on phase morphology, thermo-mechanical analysis, and melt rheological analysis revealed that the blends formed a physically crosslinked semicrystalline polymer network, and the polyaniline nanodomains dispersed in the semicrystalline POE matrix acted as photothermal agents, as well as crosslinked points. Due to photothermal heating by the PANI nanoparticles, the crosslinked blends exhibited shape memory effects under NIR light irradiation. This novel shape memory elastomer may have potential applications, such as in remotely and spatially controllable soft actuators and grippers.

**Author Contributions:** Conceptualization, Y.-W.C.; data curation, M.-S.H. and T.-H.K.; writing—original draft preparation, Y.-W.C. and K.S.J.; writing—review and editing, Y.-W.C.; project administration, Y.-W.C. All authors have read and agreed to the published version of the manuscript.

**Funding:** This research was funded by the Ministry of Trade, Industry, and Energy, Republic of Korea. (Project # 20013166).

**Informed Consent Statement:** Informed consent was obtained from all subjects involved in the study.

**Data Availability Statement:** The data presented in this study are available on request from the corresponding author.

**Conflicts of Interest:** The authors declare no conflict of interest.

## References

1. Liu, C.; Qin, H.; Mather, P.T. Review of progress in shape-memory polymers. *J. Mater. Chem.* **2007**, *17*, 1543–1558. [CrossRef]
2. Hu, J.; Zhu, Y.; Huang, H.; Lu, J. Recent advances in shape-memory polymers: Structure, mechanism, functionality, modelling and applications. *Prog. Polym. Sci.* **2012**, *37*, 1720–1763. [CrossRef]
3. Behl, M.; Razzaq, M.Y.; Lendlein, A. Multifunctional shape-memory polymers. *Adv. Mater.* **2010**, *22*, 3388–3410. [CrossRef]
4. Zhao, Q.; Qi, H.J.; Xie, T. Recent progress in shape memory polymer: New behavior, enabling materials, and mechanistic understanding. *Prog. Polym. Sci.* **2014**, *49–50*, 79–120. [CrossRef]
5. Serrano, M.C.; Ameer, G.A. Recent insights into the biomedical applications of shape-memory polymers. *Macromol. Biosci.* **2012**, *12*, 1156–1171. [CrossRef]
6. Lai, H.; Shang, Y.; Cheng, Z.; Lv, T.; Zhang, E.; Zhang, D.; Wang, J.; Liu, Y. Control of tip nanostructure on superhydrophobic shape memory arrays toward reversibly adjusting water adhesion. *Adv. Comp. Hybrid. Mater.* **2019**, *2*, 753–762. [CrossRef]
7. Wang, W.; Shen, R.; Cui, H.; Cui, Z.; Liu, Y. Two-stage reactive shape memory thiol–epoxy–acrylate system and application in 3D structure design. *Adv. Comp. Hybrid. Mater.* **2020**, *3*, 41–48. [CrossRef]
8. Basak, S. Redesigning the modern applied medical sciences and engineering with shape memory polymers. *Adv. Comp. Hybrid. Mater.* **2021**, *4*, 223–234. [CrossRef]
9. Gao, H.; Li, J.; Liu, Y.; Leng, J. Shape memory polymer solar cells. *Adv. Comp. Hybrid. Mater.* **2021**. [CrossRef]
10. Herath, M.; Epaarchchi, J.; Islam, M.; Fang, L.; Leng, J. Light activated shape memory polymers and composites: A review. *Eur. Polym. J.* **2020**, *136*, 109912. [CrossRef]
11. Zhang, H.; Xia, H.; Zhao, Y. Optically triggered and spatially controllable shape-memory polymer–gold nanoparticle composite materials. *J. Mater. Chem.* **2012**, *22*, 845–849. [CrossRef]
12. Yenpech, N.; Intasanta, V.; Chirachanchai, S. Laser-triggered shape memory based on thermoplastic and thermoset matrices with silver nanoparticle. *Polymer* **2019**, *182*, 121792. [CrossRef]
13. Ishii, S.; Uto, K.; Niiyama, E.; Ebara, M.; Nagao, T. Hybridizing poly(ε-caprolactone) and plasmonic titanium nitride nanoparticles for broadband photoresponsive shape memory films. *ACS Appl. Mater. Interfaces* **2016**, *8*, 5634–5640. [CrossRef] [PubMed]
14. Yang, L.; Tong, R.; Wang, Z.; Xia, H. Polydopamine paticle-filled shape-memory polyurethane composites with fast near-infrared light responsibility. *Chem. Phys. Chem.* **2018**, *19*, 2052–2057. [CrossRef]
15. Qian, W.; Song, Y.; Shi, D.; Dong, W.; Wang, X.; Zhang, H. Photothermal-triggered shape memory polymer prepared by cross-linking porphyrin-loaded micellar particles. *Materials* **2019**, *12*, 496. [CrossRef] [PubMed]
16. Park, J.; Kim, B. Infrared light actuated shape memory effects in crystalline polyurethane/graphene chemical hybrids. *Smart Mater. Struct.* **2014**, *23*, 025038. [CrossRef]
17. Kashif, M.; Chang, Y.-W. Supramolecular hydrogen-bonded polyolefin elastomer/modified graphene nanocomposites with near infrared responsive shape memory and healing properties. *Eur. Polym. J.* **2015**, *66*, 273–281. [CrossRef]
18. Koerner, H.; Price, G.; Pearce, N.; Alexander, M.; Vaia, R.A. Remotely actuated polymer nanocomposites-stress-recovery of carbon-nanotube-filled thermoplastic elastomers. *Nat. Mater.* **2004**, *3*, 115–120. [CrossRef]
19. Wu, S.; Li, W.; Sun, Y.; Pang, X.; Zhang, X.; Zhuang, J.; Zhang, H.; Hu, C.; Lei, B.; Liu, Y. Facile fabrication of a CD/PVA composite polymer to access light-responsive shape-memory effects. *J. Mater. Chem. C* **2020**, *8*, 8935–8941. [CrossRef]
20. Chen, M.; Fang, X.; Tang, S.; Zheng, N. Polypyrrole nanoparticles for high-performance in vivo near-infrared photothermal cancer therapy. *Chem. Comm.* **2012**, *48*, 8934–8936. [CrossRef]
21. Li, L.; Liang, K.; Hua, Z.; Zou, M.; Chen, K.; Wang, W. A green route to water-soluble polyaniline for photothermal therapy catalyzed by iron phosphates peroxidase mimic. *Polym. Chem.* **2015**, *6*, 2290–2296. [CrossRef]
22. Zhou, J.; Lu, Z.; Zhu, X.; Wang, X.; Liao, Y.; Ma, Z.; Li, F. NIR photothermal therapy using polyaniline nanoparticles. *Biomaterials* **2013**, *34*, 9584–9592. [CrossRef] [PubMed]
23. Chen, P.; Ma, Y.; Zheng, Z.; Wu, C.; Wang, Y.; Liang, G. Facile syntheses of conjugated polymers for photothermal tumour therapy. *Nat. Commun.* **2019**, *10*, 1192. [CrossRef] [PubMed]
24. Bai, Y.; Zhang, J.; Chen, X. A thermal-, water-, and near-infrared light-induced shape memory composite based on polyvinyl alcohol and polyaniline fibers. *ACS Appl. Mater. Interfaces* **2018**, *10*, 14017–14025. [CrossRef] [PubMed]
25. Dong, Y.; Geng, C.; Liu, C.; Gao, J.; Zhou, Q. Near-infrared light photothermally induced shape memory and self-healing effects of epoxy resin coating with polyaniline nanofibers. *Synth. Met.* **2020**, *266*, 116417. [CrossRef]
26. Zhang, Y.; Zhou, S.; Chong, K.C.; Wang, S.; Liu, B. Near-infrared light-induced shape memory, self-healable and anti-bacterial elastomers prepared by incorporation of a diketopyrrolopyrrole-based conjugated polymer. *Mater. Chem. Front.* **2019**, *3*, 836. [CrossRef]
27. Shimizu, S.; Saitoh, T.; Uzawa, M.; Yuasa, M.; Yano, K.; Maruyama, T.; Watanabe, K. Synthesis and applications of sulfonated polyaniline. *Synth. Met.* **1997**, *85*, 1337–1338. [CrossRef]
28. Roy, B.C.; Gupta, M.D.; Bhowmik, L.; Ray, J.K. Studies on water soluble conducting polymer aniline initiated polymerization of m-aminobenzene sulfonic acid. *Synth. Met.* **1999**, *100*, 233–236. [CrossRef]
29. Liao, Y.; Strong, V.; Chian, W.; Wang, X.; Li, X.G.; Kaner, R.B. Sulfonated polyaniline nanostructures synthesized by rapid initiated copolymerization with controllable morphology, size and electrical properties. *Macromolecules* **2012**, *45*, 1570–1579. [CrossRef]
30. Kashif, M.; Chang, Y.W. Preparation of supramolecular thermally repairable elastomer by crosslinking of maleated polyethylene-octene elastomer with 3-amino-1,2,4-triazole. *Polym. Int.* **2014**, *63*, 1936–1948. [CrossRef]

31. Wu, C.S. Preparation and characterization of aromatic polyester/polyaniline composite and its improved counterpart. *eXPRESS Polym. Lett.* **2012**, *6*, 465–473. [CrossRef]
32. Heydaria, M.; Moghadama, P.N.; Fareghia, A.R.; Bahramb, M.; Movagharnezhad, N. Synthesis of water-soluble conductive copolymer based on polyaniline. *Polym. Adv. Technol.* **2015**, *26*, 250–254. [CrossRef]
33. Pernot, H.; Baumert, M.; Court, F.; Leibler, L. Design and properties of co-continuous nanostructured polymers by reactive blending. *Nat. Mater.* **2002**, *1*, 54–58. [CrossRef]
34. Choi, M.C.; Jung, J.Y.; Chang, Y.W. Shape memory thermoplastic elastomer from maleated polyolefin elastomer and nylon 12 blends. *Polym. Bull.* **2014**, *71*, 625–635. [CrossRef]
35. Wu, F.; Misra, M.; Mohanty, A.K. Novel tunable super-tough materials from biodegradable polymer blends: Nano-structuring through reactive extrusion. *RSC Adv.* **2019**, *9*, 2836–2847. [CrossRef]
36. Chang, Y.W.; Eom, J.P.; Kim, J.; Kim, H.T.; Kim, D.K. Preparation and characterization of shape memory polymer networks based on carboxylated telechelic poly(ε-caprolactone)/epoxidized natural rubber blends. *J. Ind. Eng. Chem.* **2010**, *16*, 256–260. [CrossRef]
37. Liu, C.; Chun, S.B.; Mather, P.T.; Zheng, L.; Haley, E.H.; Coughlin, E.B. Chemically cross-linked polycyclooctene: Synthesis, characterization, and shape memory behavior. *Macromolecules* **2002**, *35*, 9868–9874. [CrossRef]
38. Sung, Y.T.; Kim, C.K.; Lee, H.S.; Kim, J.S.; Yoon, H.G.; Kim, W.N. Effects of crystallinity and crosslinking on the thermal and rheological properties of ethylene vinyl acetate copolymer. *Polymer* **2005**, *46*, 11844–11848. [CrossRef]
39. Sasmal, A.; Sahoo, D.; Nanda, R.; Nayak, P.; Nayak, P.L.; Mishra, J.K.; Chang, Y.W.; Yoon, J.Y. Biodegradable nanocomposites from maleated polycaprolactone/soy protein isolate blend with organoclay: Preparation, characterization, and properties. *Polym. Comp.* **2009**, *30*, 708–714. [CrossRef]

Article

# Enhancement of the Processability and Properties of Nylon 6 by Blending with Polyketone

Tao Zhang and Ho-Jong Kang *

Department of Polymer Science and Engineering, Dankook University, 152 Jukjeon-ro, Suji-gu, Yongin-si 16889, Gyeonggi-do, Korea; taozhang1214@gmail.com
* Correspondence: hjkang@dankook.ac.kr

**Abstract:** Polyketones (PKs) having strong hydrogen bonding properties and a chain extender are used as additives in the melt processing of nylon 6 (PA6). Their effect on the chain structure and properties of PA6 is studied to enhance the processability of PA6 in melt processing. The addition of the chain extender to PA6 increases the melt viscosity by forming branches on the backbone. The addition of PKs results in an additional increase in viscosity through the hydrogen bonding between N–H of PA6 and C=O of PK. The change in the N–H bond FT-IR peak of PA6 and the swelling data of the PA6/PK blend containing a chain extender, styrene maleic anhydride copolymer (ADR), suggest that incorporation of chain extender and PK in the melt processing of PA6 results in physical crosslinks through hydrogen bonding between the branched PA6 formed by the addition of chain extender and PK chains. This change in the chain structure of PA6 not only increases the melt strength of PA6 but also increases randomness resulting in decreased crystallinity.

**Keywords:** nylon 6; polyketone; chain extender; hydrogen bonding; chain branching; chain crosslinking; melt viscosity

**Citation:** Zhang, T.; Kang, H.-J. Enhancement of the Processability and Properties of Nylon 6 by Blending with Polyketone. *Polymers* **2021**, *13*, 3403. https://doi.org/10.3390/polym13193403

Academic Editor: Kwang-Jea Kim

Received: 28 August 2021
Accepted: 28 September 2021
Published: 3 October 2021

**Publisher's Note:** MDPI stays neutral with regard to jurisdictional claims in published maps and institutional affiliations.

**Copyright:** © 2021 by the authors. Licensee MDPI, Basel, Switzerland. This article is an open access article distributed under the terms and conditions of the Creative Commons Attribution (CC BY) license (https:// creativecommons.org/licenses/by/ 4.0/).

## 1. Introduction

Polyamide is the most widely used thermoplastic polymer among engineering plastics due to its superior mechanical properties and chemical resistance [1–3]. PA6 has generally been used as a textile material substituting natural fibers [4,5], but recently, it is being used widely for automotive parts due to its excellent impact resistance and abrasion resistance [6–8]. PA6 automotive parts are generally processed by melt processing such as extrusion or injection molding, and for thermal stability, chain extenders can be used to introduce branches and increase the molecular weight [9–11]. Diverse reinforcing materials can also be introduced to prepare nylon composites [12,13]. As the extrusion or injection molding of PA6 is carried out at a relatively high temperatures above 220 °C, chain scission by thermal degradation results in a decrease in the molecular weight [14], and thus, the melt strength and mechanical properties of the product are decreased significantly. The thermal degradation is known to be accelerated by the presence of water in PA6 [15].

To deal with the thermal degradation during melt processing, the following methods have been reported: solid-state polymerization (SSP), where the thermal degradation products can be removed by the introduction of inert gas in a vacuum state to minimize further thermal degradation [16]; and extension of the degraded chain by the introduction of chain extenders containing functional groups, which can react with the diverse end groups produced by the degradation of PA6 [17]. The most widely used chain extenders for PA6 are those that can react with the carboxyl (–COOH) or amino (–$NH_2$) groups produced by the thermal degradation of PA6 in the presence of water such as bis caprolactam [18], bislaurolactamdiaryl lactam, or bisoxazoline [19] or those containing anhydride functional groups, which can react with O=C–$NH_2$ or –CH=$CH_2$ end groups produced by thermal degradation in the absence of water, such as Joncryl ADR [20,21] or epoxy resin [22]. The addition of a chain extender increases the molecular weight of PA6 by branching or partial crosslinking of PA6 through reaction with the end groups produced in the high-temperature melt processing to result in an increase in the melt strength, thus enhancing the processability and overall properties such as mechanical properties.

PA6, poly(ethyl cyanoacrylate)(PECA), PK, etc. with polar groups such as –NH, –OH, and C=O can exhibit dipole–dipole interaction, ion–dipole interaction, and hydrogen bonding in blending. Especially, PK having –C=O groups in the main chain capable of forming strong hydrogen bonds is reported to be very compatible with PA6 in melt processing due to its hydrogen bonding with the –NH bonds of PA6 [23–26]. Although this hydrogen bonding is expected to affect the chain structure of PA6 and thus the overall properties, there is no report emphasizing this aspect. Direct modification of PK with diamines resulted in a dramatic increase of the tensile strength of PK [27]. It was also found that hydrogen bonding between PK and PECA resulted in interpenetrating networks by which wettability and morphology can be controlled [28].

In this study, enhancement of the melt strength and mechanical properties through the introduction of branches to PA6 is attempted by adding PK, thereby, enhancing the melt viscosity of PA6 in injection molding. The capability of PK to form hydrogen bonds with PA6 in the melt processing of PA6 with chain extender was investigated. The effect on the chain structure of PA6 and the resulting change in the rheological and mechanical properties are studied.

## 2. Experimental

### 2.1. Materials

The PA6 used in this study was Taekwang RV 2.10 (Seoul, Korea), with a melting point of 221 °C, specific gravity of 1.12 g/cm$^3$. The PK used as an additive was Hyosung M630A (Seoul, Korea), a terpolymer of carbon monoxide, ethylene, and propylene, with a melting point of 218 °C, MI 6 g/10 min, propylene content of 5.64%, and specific gravity of 1.24 g/cm$^3$. The chain extender, multifunctional styrene-acrylic oligomer (Joncryl ADR 4370, Qingdao, China), was purchased from BASF. Antioxidant ZIKA-1010 6683-19-8 was purchased from ZIKO (Anyang-si, Korea) and used to minimize thermal degradation in the melt processing. Tetrahydrofuran (THF, Sigma-Aldrich, Darmstadt, Germany) was used to measure the degree of swelling of PA6 and PA6/PK blend.

### 2.2. Sample Preparation and Reactive Processing

PA6 was vacuum dried at 80 °C for 24 h. to minimize the hydrolytic thermal degradation prior to melt blending. Haake internal mixer (Karlsruhe, Germany) was used to blend dried PA6, ADR, PK, and antioxidant ZIKA-1010 at different ratios. The antioxidant concentration was fixed at 0.2%, and 1, 3, 5 phr ADR and 1, 3, 5, 10 wt.% PK were added, respectively or together, to 40 g PA6 and blended at 220 or 260 °C for 15 min at a stirring speed of 30 rpm. The change in torque with blending time of PA6 with PK and ADR was observed. The blend was compression molded at 220 °C into a 20 mm × 20 mm mold of different thicknesses on a QMESYS QM900A (Uiwang, Korea) and quenched to 4 °C to obtain the samples. Samples (1 mm thick) were prepared for rheological testing, FT-IR and degree of swelling tests, and 0.35-mm-thick samples were prepared for tensile testing.

### 2.3. Characterization

Rotational rheometer AR2000ex (TA Instruments, New Castle, DE, USA) was used to evaluate the rheological properties of the melt. The sample was attached to a 25 mm ETC steel plate, then the complex viscosity and the loss tangent were measured at a rotation speed of 0.1–628 rad/s and 1% deformation at 220 °C.

The crystallinity of the samples ($\chi$) was calculated using the following equation with data obtained on a TA differential scanning calorimeter Q20 (TA Instruments, New Castle, DE, USA) scanning from −50 to 260 °C, at a heating rate of 10 °C/min.

$$\chi = \frac{\Delta H_m}{\Delta H^o_m} \times 100\% \tag{1}$$

where $\Delta H_m$: enthalpy of melting, $\Delta H^o{}_m$: enthalpy of melting of pure PA6 (190 J/g) [29].

Infrared spectra of the PA6 and PA6/PK blends were obtained on a Thermo Scientific Nicolet iS10 FT-IR (Thermo Scientific, Waltham, MA, USA) in the ATR mode, in the range 4000–500 cm$^{-1}$ at a resolution of 4 cm$^{-1}$ and scan number of 16.

The degree of swelling in THF was measured to confirm the changes in the chain structure such as partial crosslinking of PA6 on mixing with ADR and PK. A 20 mg (Wd) sample was put in 0.6 mLTHF and sonicated at 50–70 °C for 100 min on a Hwashin Powersonic 410 (Gwangju-si, Korea), then the weight of the sample (Ww) was measured to calculate the degree of swelling (DS%) with the following equation.

$$DS(\%) = \frac{Ww - Wd}{Wd} \times 100 \qquad (2)$$

The mechanical properties were evaluated by measuring the tensile strength, Young's modulus, and elongation at break with 10 mm × 20 mm × 0.35 mm samples on a Lloyd tensile tester LR30K (LLOYD, Cleveland, OH, USA) at a crosshead speed of 10 mm/min.

## 3. Results and Discussion

The change in the torque of the mixer along with the mixing time of PA6 and PK at 220 °C is shown in Figure 1a. As can be seen in the figure, the torque increased rapidly initially when PA6 was fed into the mixer as it started to melt but stabilized within 2 min on melting. Once the torque stabilized after the addition of PK, the torque was higher compared with that of PA6 alone, which was due to the higher melt viscosity of PK. Another significant observation was that after 6 min, when PA6 and PK were substantially mixed, the torque again increased. The rapid increase in torque in the molten state suggests that reaction or mutual interaction was occurring. As PA6 and PK both have polar groups such as –NH and –C=O, we can expect hydrogen bonding between the polar groups in the PA6 domain and PK co-domain in the blend, once mixing occurs to a certain degree (Figure 1a). The schematic is shown in Figure 2a. Figure 1b shows the change in torque with time when 5 phr of the commonly used chain extender Joncryl ADR was added to PA6 prior to the addition of PK. Compared with the PA6/PK blend in Figure 1a, the torque initially did not show much difference when ADR was added prior to mixing PK; however, after a relatively short mixing time of 4 min, a drastic torque change occurred due to the interaction of the polar groups, with the increase being higher at higher PK contents. In addition, the time at which the torque started to increase decreased with increase in the PK content. This shows that the epoxy groups of the chain extender ADR and –COOH or –NH$_2$ groups of PA6 reacted to form branches on PA6 (Figure 2b), and then physical crosslinks were formed between the branched PA6 and PK chains through hydrogen bonding interaction (Figure 2c).

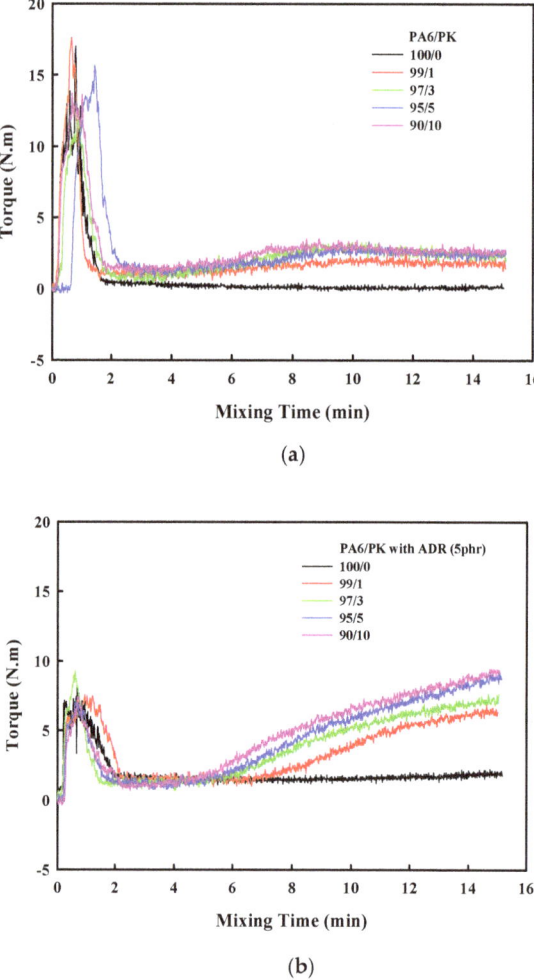

**Figure 1.** Melt torque as a function of mixing time for PA6/PK blends; (**a**) without chain extender; (**b**) with chain extender(5 phr).

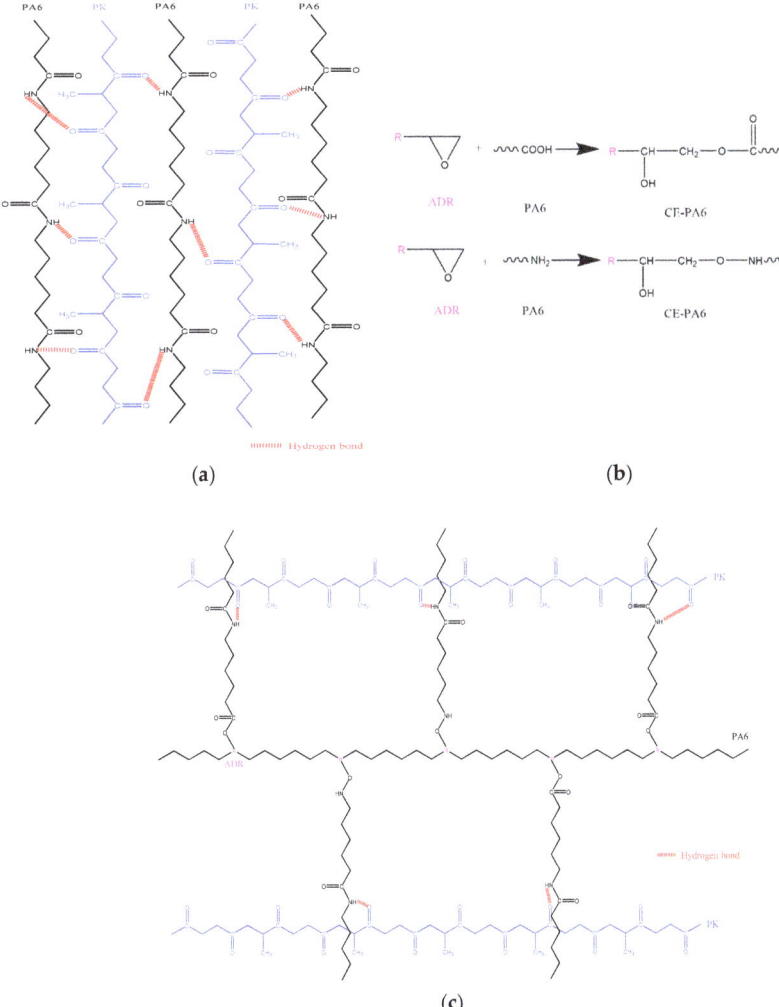

**Figure 2.** Schematics of chain extension: (**a**) PA6/PK blends; (**b**) PA6 with ADR; (**c**) PA6/PK blends with ADR.

Figure 3a shows the effect of ADR content on the torque change in PA6 and PA6/PK (90/10) blend during the mixing process. In the case of PA6, a slight increase in the torque was observed with the increase in the ADR content due to chain extension, but a drastic increase in the torque was not observed with mixing time. However, in the case of the PA6/PK blend, a drastic increase in the torque was observed with increase in the ADR content. This suggests that with increase in the ADR content more branches occurred on PA6 due to chain extension, and the branched PA6 chains affected the hydrogen bonding with PK to change the resulting structure, that is, the structure was not the linear PA6 structure in Figure 2a, but rather the structure with physical crosslinks shown in Figure 2c. The effects of ADR content and mixing temperature on the change in torque in PA6 and PA6/PK blends are shown in Figure 3. Increase in the ADR content in PA6 increased the content of reactive ADR epoxy groups to increase the number of branches resulting in overall increase in the torque. However, in the case of PA6/PK blends, increase in the ADR content resulted in a drastic increase in the torque due to an increase in hydrogen bonding between the branches on PA6 and PK. The effect of mixing temperature on the change in torque was negligible in the case of PA6 due to its relatively low melt viscosity, but when ADR was added, the torque increased with increase in the mixing temperature, and when ADR and PK were added, the torque increase became more drastic and the time at which the torque increases became shorter. This suggests that the increase in the mixing temperature resulted in an increase in the formation of end groups with which the added ADR could react to form branches and in a drastic increase in the torque through hydrogen bonding between the branched PA6 chains and PK.

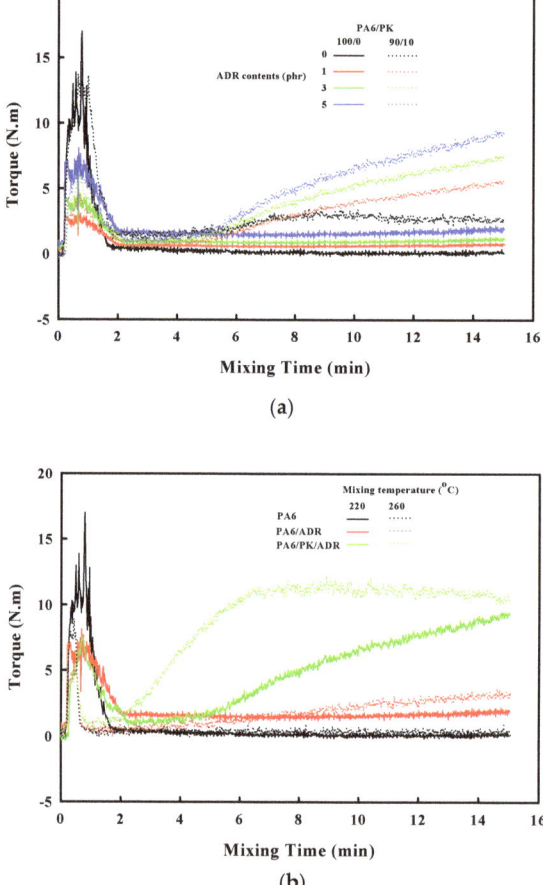

**Figure 3.** Effect of mixing time on the melt torque of PA an PA6/PK(90/10) blend (**a**) at different ADR contents and (**b**) processed at different mixing temperatures.

The change in the FT-IR spectrum with addition of the chain extender ADR in PA6 and PA6/PK 90/10 blend is shown in Figure 4, and the change in the maximum absorption of the respective peaks with the amount of ADR added is shown in Figure 5. The change in the maximum absorption of the characteristic FT-IR peaks of the amine N-H stretching peak at 3294 cm$^{-1}$ and the bending peak at 1538 cm$^{-1}$ of PA6 with ADR content are shown in Figure 5a,b, and the change in the maximum absorption of the FT-IR peak of the NHC=O bond at 1634 cm$^{-1}$ and C=O bond at 1699 cm$^{-1}$ in PK and PA6/PK blends are shown Figure 5c,d, respectively. There was no change in the maximum absorption of the C=O bond in PK with ADR content, while the maximum absorption of the –NH bond and

NHC=O bond of PA6 decrease with the addition of PK and ADR. This shows again that the addition of ADR results in a change in the chain structure from the formation of branches and that a change in the chain structure due to hydrogen bonding between the polar C=O of PK and the NH– and –NHC=O of PA6 is occurring simultaneously.

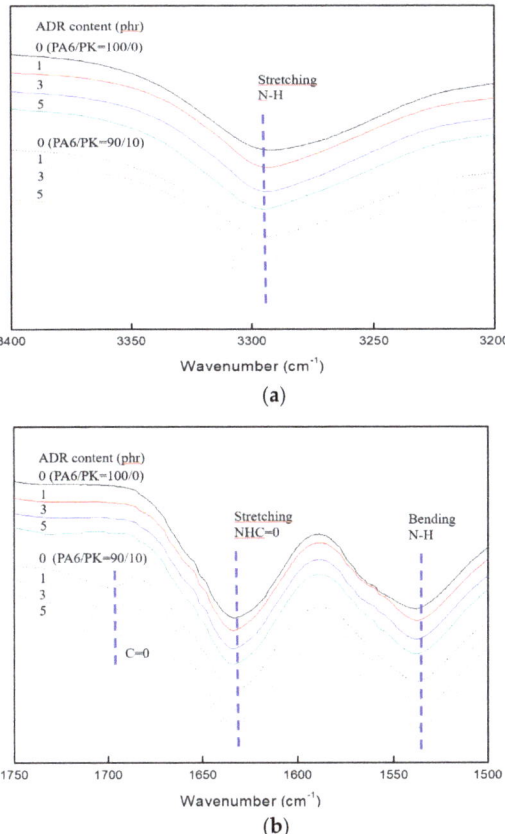

**Figure 4.** FTIR spectra of PA6 and PA6/PK blend (90/10) of different ADR contents: (**a**) N–H stretching (3294 cm$^{-1}$); (**b**) C=O (1699 cm$^{-1}$), NHC=O stretching (1634 cm$^{-1}$), NHC=O stretching (1634 cm$^{-1}$), N–H bending (1538 cm$^{-1}$).

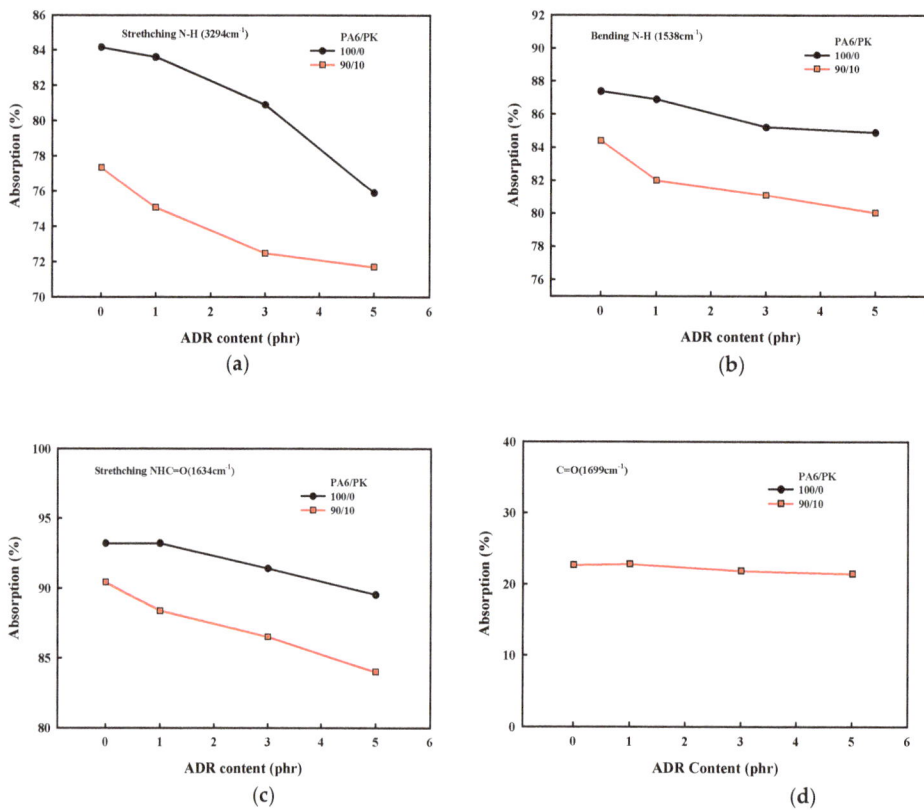

**Figure 5.** Effect of ADR content on the maximum absorption of the peaks due to amino and carbonyl groups in PA6 and PA6/PK blends: (**a**) N–H stretching (3294 cm$^{-1}$); (**b**) N–H bending (1538 cm$^{-1}$); (**c**) NHC=O stretching (1634 cm$^{-1}$); (**d**) C=O (1699 cm$^{-1}$).

To confirm the change in the chain structure due to hydrogen bonding between PA6 and PK, the change in the degree of swelling of the PA6/PK blend THF at 50 °C with ADR content is shown in Figure 6. The PA6 and PA6/PK blend dissolved completely in THF in the absence of ADR, and PA6 dissolved in THF even when ADR was present, but when ADR was added to the PA6/PK blends, they did not dissolve in THF but just swelled. The degree of swelling of the PA6/PK blend increased with the ADR content and amount of PK in the blend. This result suggests that the change in the chain structure on addition of ADR to PA6 and to PA6/PK blend is different. That is, chain extension of PA6 with ADR resulted in branched PA6 (Figure 2b), which is soluble in THF, while the addition of ADR to PA6/PK blends resulted in physical crosslinking between the branched PA6 chains and PK through hydrogen bonding (Figure 2c), and thus they did not dissolve in THF but swelled. The increase in ADR content increased the degree of branching of PA6, and thus, the amount of hydrogen bonding with PK to increase the degree of swelling. To observe the effect of dissolution temperature on the solubility of PA6 in THF, the change in the degree of swelling with solvent temperature is shown in Figure 7, where the sample was the PA6/PK 90/10 blend in which physical crosslinking through hydrogen bonding was expected. If the crosslinking was chemical, the degree of swelling should increase with solvent temperature; however, the degree of swelling decreased and there was also a change in the shape of the sample suggesting an increase in the solubility results in partial dissolution of the ADR containing PA6/PK blend. This substantiates the observation that the crosslinking between branched PA6 and PK in the presence of ADR was not chemical but physical crosslinking.

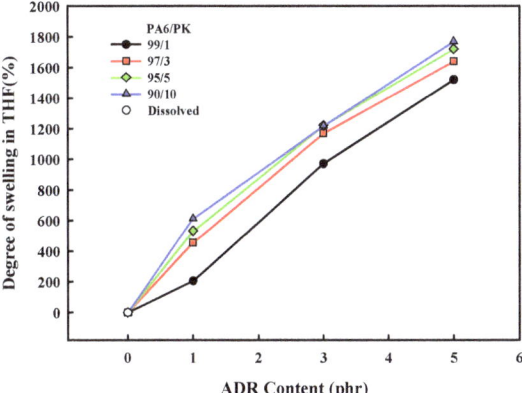

**Figure 6.** Effect of chain extender, ADR content on the degree of swelling of PA6/PK blends swollen in THF at 50 °C for 100 min.

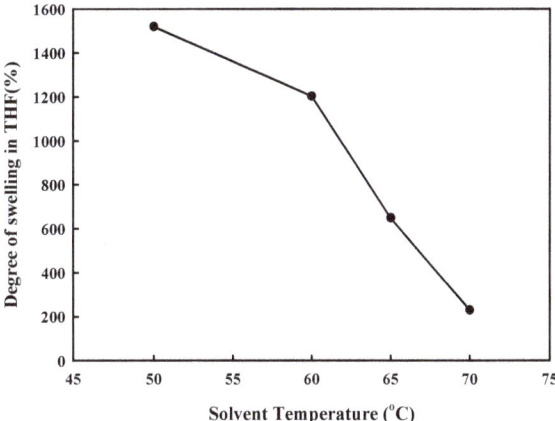

**Figure 7.** Effect of solvent temperature on the degree of swelling of PA6/PK(90/10) blend with ADR in THF for 100 min.

The dynamic rheological properties are known to be greatly affected by branching and crosslinking [30]. The complex viscosity and the loss tangent of PA6 and PA6/PK blend shown in Figure 8 show that both exhibited a typical non-Newtonian shear thinning behavior and that the addition of ADR resulted in an increase in the viscosity with the shear thinning behavior becoming more pronounced with increase in the ADR content. The effect of chain branching on the shear viscoelasticity is generally known to be dependent on the length of the branch [31]. Short branches cause a small change in the viscosity and a small change in viscosity with shear rate as they are within the entanglement radius of the melt, while long branches protrude out of the entanglement radius to increase the viscosity and accentuate the decrease in the viscosity with increase in shear rate. The loss tangent shown in Figure 8b also decreased with the formation of long chain branches in the presence of chain extender. The loss tangent has an inverse relationship with the melt strength, which affects the melt processing properties [32]. That is, the melt processability of PA6 was enhanced by the addition of ADR due to the formation of long chain branches.

In the case of PA6/PK blends, the complex viscosity increased due to the relatively high viscosity of PK and the suggested hydrogen bonding between PA6 and PK, exhibiting a drastic increase with addition of ADR, and the shear thinning behavior characteristic of non-Newtonian fluids also showed a more drastic change compared with when only ADR was added to PA6. This shows that the change in chain structure due to ADR was completely different in the case of PA6 and PA6/PK blends as shown in Figure 2. As seen in Figures 6 and 7, contrary to the formation of long chain branches with the addition of ADR to PA6, the hydrogen bonding between branched PA6 and PK formed physical crosslinks, which caused a drastic increase in the viscosity and shear thinning behavior in the case of PA/PK blends. The change in the viscoelastic properties with these changes in the chain structure resulted in a significant decrease in the loss tangent at low shear rates, as can be seen in Figure 8b. Thus, minimization of the decrease in the molecular weight through chain extension and the formation of physical crosslinks between branched PA6 and PK may also solve the low melt viscosity drawback in PA6 melt processing. The effect of mixing temperature on the rheological properties of PA6/PK blends shown in Figure 9. PA6 did not exhibit a significant change in the viscosity with the change in mixing temperature from 220 to 260 °C, suggesting that there was not a significant change in the extent of thermal degradation at these temperatures. When ADR was added to PA6, an increase in viscosity occurred. This suggests that although there was not a significant decrease in the molecular weight, more end groups were formed that reacted with ADR to form branches increasing the viscosity. However, the loss tangent increased, suggesting there was a change in the length of the branches with change in mixing temperature. In the case of PA6/PK blend with added ADR, the mixing temperature did not have an effect on the melt viscosity or the loss tangent, suggesting that the change in the chain structure with change in mixing temperature did not have a significant effect on their rheological behavior. This suggests that the physical crosslinking through hydrogen bonding had a greater effect on the rheological properties compared with that of the branching of PA6.

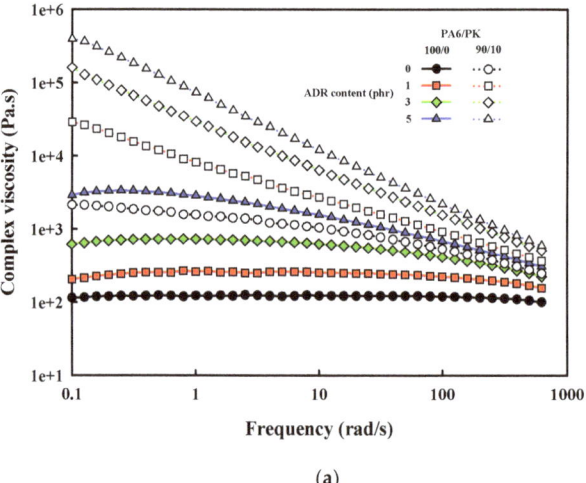

(a)

Figure 8. Cont.

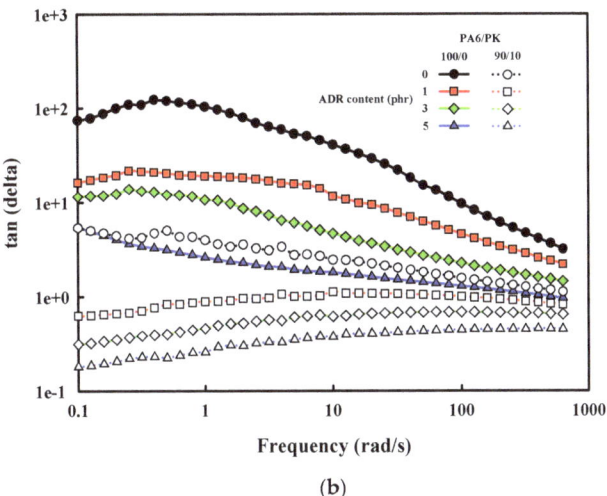

(b)

**Figure 8.** Rheological properties of PA6 and PA6/PK blend with chain extender, ADR: (**a**) complex viscosity; (**b**) loss tangent.

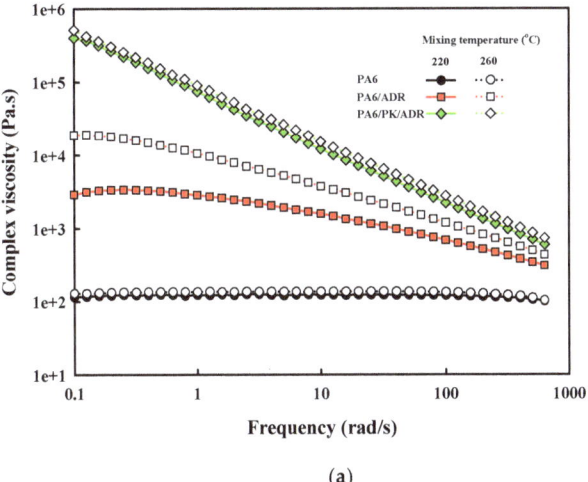

(a)

**Figure 9.** *Cont.*

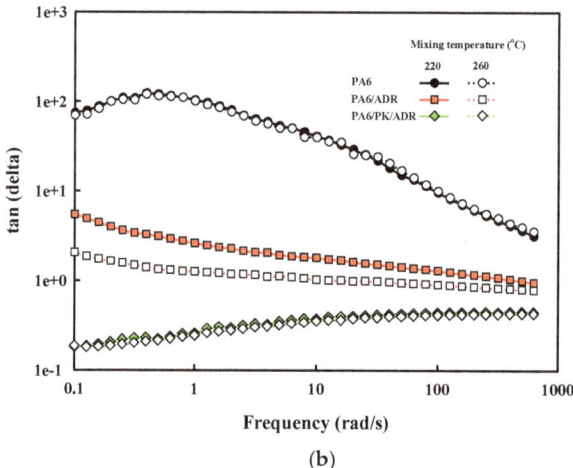

**Figure 9.** Effect of mixing temperature on the rheological properties of PA6 and PA6/PK blend with chain extender, ADR: (**a**) complex viscosity; (**b**) loss tangent.

The change in the crystallinity with the addition of ADR to PA6 and PA6/PK blends is shown in Figure 10. The crystallinity of PA6 decreased on blending with PK and that of PA6 and PA6/PK blends all decreased with the addition of ADR. The decrease in crystallinity with addition of ADR was due to the change in the chain structure caused by the chain extension with ADR. As confirmed, the long chain branching and the physical crosslinking effected by addition of ADR and PK decreased the regularity of the chain and hindered the crystallization of nylon 6 to decrease its crystallinity. The greater decrease in the crystallinity on addition of ADR to PA6/PK blends compared with PA6 suggests that physical crosslinking with addition of PK decreased the crystallinity to a greater extent compared with long branching formed by chain extension with ADR. The crystallinity also decreased when mixed at higher temperature and with an increase in the ADR content, suggesting that the degree of branching of PA6 with ADR also influenced the crystallization of PA6. The changes in the mechanical properties of PA6 and PA6/PK blends with ADR content in Figure 11 show that the mechanical properties were enhanced by the addition of ADR and PK. The change in chain structure by the introduction of long branches and physical crosslinks through the addition of ADR increased the melt viscosity to enhance the melt processability, which also increased the chain orientation. Increase in chain orientation is known to increase the physical properties of polymers such as tensile strength and elastic modulus. Especially, physical crosslinking exhibits a greater increase in the elongation at break compared with long chain branching, as observed in previous figures.

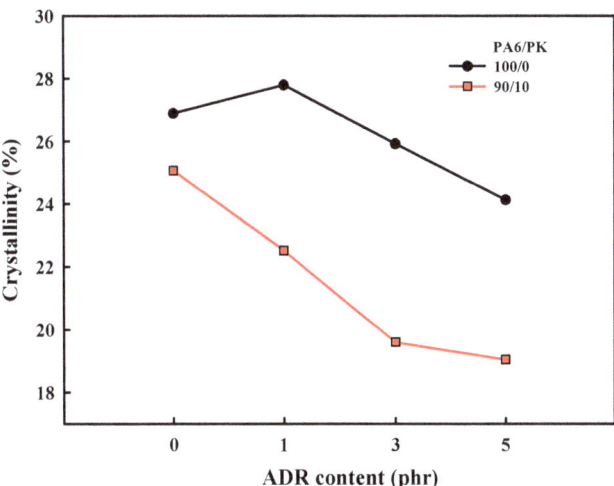

**Figure 10.** Effect of chain extender, ADR, content on the crystallinity of PA6 and PA6/PK blends with chain extender, reflecting the influence of chain branching.

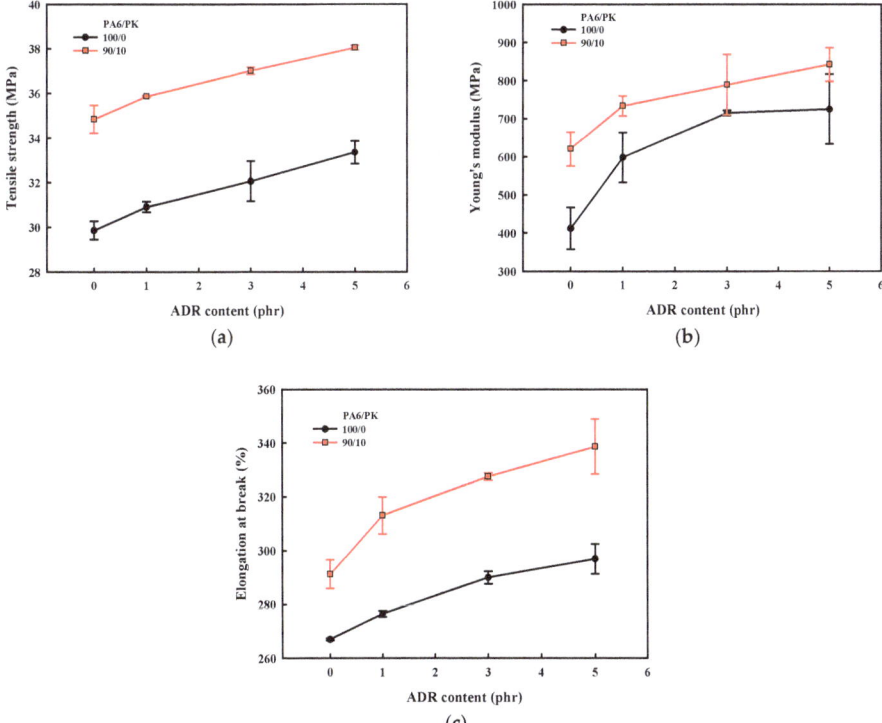

**Figure 11.** Effect of chain extender, ADR, content on the mechanical properties of PA6 and PA6/PK blends with chain extender, reflecting the influence of chain branching: (**a**) tensile strength; (**b**) Young's modulus; (**c**) elongation at break.

## 4. Conclusions

Chain extenders used widely in the melt processing of polymers and PK capable of hydrogen bonding were added in the melt processing of PA6, and their effects on the chain structure and properties of PA6 were studied. Long chain branching was formed on PA6 through the addition of the chain extender ADR, and physical crosslinking was introduced by further addition of PK, which formed hydrogen bonds with long chain branched PA6. These changes in the chain structure increased the melt viscosity of PA6 and accentuated the shear thinning behavior to enhance the melt processability. Especially, the physical crosslinking introduced by the addition of PK was effective in the enhancement of the melt processability of PA6 and also resulted in a decrease in the crystallinity and an increase in the physical properties.

**Author Contributions:** Conceptualization, H.-J.K.; investigation, T.Z.; writing—original draft, T.Z. and H.-J.K. All authors have read and agreed to the published version of the manuscript.

**Funding:** This research was funded by Gyeonggi-do through the Gyeonggi-do Regional Research Center (GRRC) Program (Project: Development of functional microcomposite materials for precision molding of flexible materials, GRRC Dankook 2016-B02).

**Institutional Review Board Statement:** Not applicable.

**Informed Consent Statement:** Informed consent was obtained from all subjects involved in the study.

**Data Availability Statement:** Data are in the authors' possession.

**Conflicts of Interest:** The authors declare no conflict of interest.

## References

1. Wang, Y.; Hou, D.F.; Ke, K.; Huang, Y.H.; Yan, Y.; Yang, W.; Yin, B.; Yang, M.B. Chemical-resistant polyamide 6/polyketone composites with gradient encapsulation structure: An insight into the formation mechanism. *Polymer* **2021**, *212*, 123173–123182. [CrossRef]
2. Ou, X.; Lu, X.; Chen, S.; Lu, Q. Thermal conductive hybrid polyimide with ultrahigh heat resistance, excellent mechanical properties and low coefficient of thermal expansion. *Eur. Polym. J.* **2020**, *122*, 109368–109375. [CrossRef]
3. Kashiwagi, T.; Harris, R.H., Jr.; Zhang, X.; Briber, R.M.; Cipriano, B.H.; Raghavan, S.R.; Awad, W.H.; Shields, J.R. Flame retardant mechanism of polyamide 6–clay nanocomposites. *Polymer* **2004**, *45*, 881–891. [CrossRef]
4. Saba, N.; Tahir, P.M.; Jawaid, M. A Review on Potentiality of Nano Filler/Natural Fiber Filled Polymer Hybrid Composites. *Polymers* **2014**, *6*, 2247–2273. [CrossRef]
5. Unterweger, C.; Bruggemann, O.; Furst, C. Synthetic Fibers and Thermoplastic Short-Fiber-Reinforced Polymers: Properties and Characterization. *Polym. Compos.* **2014**, *35*, 227–236. [CrossRef]
6. Nguyen-Tran, H.D.; Hoang, V.T.; Do, V.T.; Chun, D.M.; Yum, Y.J. Effect of Multiwalled Carbon Nanotubes on the Mechanical Properties of Carbon Fiber-Reinforced, Polyamide-6/Polypropylene Composites for Lightweight Automotive Parts. *Materials* **2018**, *11*, 429. [CrossRef] [PubMed]
7. Sonsino, C.M.; Moosbrugger, E. Fatigue design of highly loaded short-glass-fiber reinforced polyamide parts in engine compartments. *Int. J. Fatigue* **2008**, *30*, 1279–1288. [CrossRef]
8. Do, V.T.; Nguyen-Tran, H.D.; Chun, D.M. Effect of polypropylene on the mechanical properties and water absorption of carbon-fiber-reinforced-polyamide-6/polypropylene composite. *Compos. Struct.* **2016**, *150*, 240–245. [CrossRef]
9. Cai, J.; Liu, Z.; Cao, B.; Guan, X.; Liu, S.; Zhao, J. Simultaneous Improvement of the Processability and Mechanical Properties of Polyamide-6 by Chain Extension in Extrusion. *Ind. Eng. Chem. Res.* **2020**, *59*, 14334–14343. [CrossRef]
10. Xu, M.; Lu, J.; Zhao, J.; Wei, L.; Liu, T.; Zhao, L.; Park, C.B. Rheological and foaming behaviors of long-chain branched polyamide 6 with controlled branch length. *Polymer* **2021**, *224*, 123730–123739. [CrossRef]
11. Dorr, D.; Kuhn, U.; Altstadt, V. Rheological Study of Gelation and Crosslinking in Chemical Modified Polyamide 12 Using a Multiwave Technique. *Polymers* **2020**, *12*, 855. [CrossRef] [PubMed]
12. Satheeskumar, S.; Kanagaraj, G. Experimental investigation on tribological behaviours of PA6, PA6-reinforced $Al_2O_3$ and PA6-reinforced graphite polymer composites. *Bull. Mater. Sci.* **2016**, *39*, 1467–1481. [CrossRef]
13. Cai, J.; Wirasaputra, A.; Zhu, Y.; Liu, S.; Zhou, Y.; Zhang, C.; Zhao, J. The flame retardancy and rheological properties of PA6/MCA modified by DOPO-based chain extender. *RSC Adv.* **2017**, *7*, 19593–19603. [CrossRef]
14. Tuna, B.; Benkreira, H. Reactive Extrusion of Polyamide 6 Using a Novel Chain Extender. *Polym. Eng. Sci.* **2019**, *59*, E25–E31. [CrossRef]
15. Tuna, B.; Benkreira, H. Chain Extension of Recycled PA6. *Polym. Eng. Sci.* **2018**, *58*, 1037–1042. [CrossRef]

16. Cakir, S.; Eriksson, M.; Martinelle, M.; Koning, C.K. Multiblock copolymers of polyamide 6 and diepoxypropyleneadipate obtained by solid state polymerization. *Eur. Polym. J.* **2016**, *79*, 13–22. [CrossRef]
17. Ozmen, S.C.; Ozkoc, V.; Serhatli, E. Thermal, mechanical and physical properties of chain extended recycled polyamide 6 via reactive extrusion: Effect of chain extender types. *Polym. Degrad. Stab.* **2019**, *162*, 76–84. [CrossRef]
18. Buccella, M.; Dorigato, A.; Pasqualini, E.; Caldara, M.; Fambri, L. Chain Extension Behavior and Thermo-Mechanical Properties of Polyamide 6 Chemically Modified with 1,10-Carbonyl-Bis-Caprolactam. *Polym. Eng. Sci.* **2014**, *54*, 158–165. [CrossRef]
19. Lu, C.; Chen, L.; Ye, R.; Cai, X. Chain Extension of Polyamide 6 Using Bisoxazoline Coupling Agents. *J. Macromol. Sci. Part B Phys.* **2008**, *47*, 986–999. [CrossRef]
20. Tuna, B.; Benkreira, H. Chain Extension of Polyamid 6/Organoclay Nanocomposites. *Polym. Eng. Sci.* **2019**, *59*, 1233–1241. [CrossRef]
21. Petrus, J.; Kucera, F.; Jancar, J. Online monitoring of reactive compatiblization of poly (lactic acid)/polyamide 6 blend with different compatibilizers. *J. Appl. Polym. Sci.* **2019**, *136*, 48005–48017. [CrossRef]
22. Wirasaputra, A.; Zhao, J.; Zhu, Y.; Liu, S.; Yuan, Y. Application of bis (glycidyloxy)phenylphosphine oxide as a chain extender for polyamide-6. *RSC Adv.* **2015**, *5*, 31878–31885. [CrossRef]
23. Jeon, I.; Lee, M.; Lee, S.W.; Jho, J.Y. Morphology and Mechanical Properties of Polyketone/Polycarbonate Blends Compatibilized with SEBS and Polyamide. *Macromol. Res.* **2019**, *27*, 827–832. [CrossRef]
24. Zhou, Y.C.; Li, S.Y.; Yang, Y.; Bao, R.Y.; Liu, Z.Y.; Yang, M.B.; Yang, W. Morphologies, interfacial interaction and mechanical performance of super-tough nanostructured PK/PA6 blends. *Polym. Test.* **2020**, *91*, 106777. [CrossRef]
25. Kim, Y.; Bae, J.W.; Lee, C.S.; Kim, S.; Jung, H.; Jho, J.Y. Morphology and Mechanical Properties of Polyketone Blended with Polyamide and Ethylene-Octene Rubber. *Macromol. Res.* **2015**, *23*, 971–976. [CrossRef]
26. Wang, Z.; Tong, X.; Yang, J.C.; Wang, X.J.; Zhang, M.L.; Zhang, G.; Long, S.R.; Yang, J. Improved strength and toughness of semi-aromatic polyamide 6T-co-6 (PA6T/6)/GO composites via in situ polymerization. *Compos. Sci. Technol.* **2019**, *175*, 6–17. [CrossRef]
27. Zhou, Y.C.; Zhou, L.; Feng, C.P.; Wu, X.T.; Bao, R.Y.; Liu, Z.Y.; Yang, M.B.; Yang, W. Direct modification of polyketone resin for anion exchange membrane of alkaline fuel cells. *J. Colloid Interface Sci.* **2019**, *556*, 420–431. [CrossRef]
28. Attanasio, A.; Bayer, I.S.; Ruffilli, R.; Ayadi, F.; Athanassiou, A. Surprising high hydrophobicity of polymer networks from hydrophilic components. *Appl. Mater. Interfaces* **2013**, *5*, 5717–5726. [CrossRef] [PubMed]
29. Han, S.; Jiang, C.; Yu, K.; Mi, J.; Chen, S.; Wang, X. Influence of crystallization on microcellular foaming behavior of polyamide 6 in a supercritical $CO_2$-assisted route. *J. Appl. Polym. Sci.* **2020**, *137*, 49183–49192. [CrossRef]
30. Xu, M.; Yan, H.; He, Q.; Wan, C.; Liu, T.; Zhao, L.; Park, C.B. Chain extension of polyamide 6 using multifunctional chain extenders and reactive extrusion for melt foaming. *Eur. Polym. J.* **2017**, *96*, 210–220. [CrossRef]
31. Xu, M.; Chen, J.; Qiao, Y.; Wei, L.; Liu, T.; Park, C.B. Toughening mechanism of long chain branched polyamide 6. *Mater. Des.* **2020**, *196*, 109173–109180. [CrossRef]
32. Xu, M.; Chen, Y.C.; Liu, T.; Zhao, L.; Park, C.B. Determination of modified polyamide 6's foaming windows by bubble growth simulations based on rheological measurements. *J. Appl. Polym. Sci.* **2019**, *136*, 48138–48141. [CrossRef]

*Article*

# Phlogopite-Reinforced Natural Rubber (NR)/Ethylene-Propylene-Diene Monomer Rubber (EPDM) Composites with Aminosilane Compatibilizer

Sung-Hun Lee †, Su-Yeol Park †, Kyung-Ho Chung * and Keon-Soo Jang *

Department of Polymer Engineering, School of Chemical and Materials Engineering, The University of Suwon, Hwaseong 18323, Korea; atlasba@naver.com (S.-H.L.); tnduf3971@naver.com (S.-Y.P.)
* Correspondence: khchung@suwon.ac.kr (K.-H.C.); ksjang@suwon.ac.kr (K.-S.J.)
† These authors (S.-H. Lee & S.-Y. Park) contributed equally: Co-1st authors.

**Abstract:** Rubber compounding with two or more components has been extensively employed to improve various properties. In particular, natural rubber (NR)/ethylene-propylene-diene monomer rubber (EPDM) blends have found use in tire and automotive parts. Diverse fillers have been applied to NR/EPDM blends to enhance their mechanical properties. In this study, a new class of mineral filler, phlogopite, was incorporated into an NR/EPDM blend to examine the mechanical, curing, elastic, and morphological properties of the resulting material. The combination of aminoethylaminopropyltrimethoxysilane (AEAPS) and stearic acid (SA) compatibilized the NR/EPDM/phlogopite composite, further improving various properties. The enhanced properties were compared with those of NR/EPDM/fillers composed of silica or carbon black (CB). Compared with the NR/EPDM/silica composite, the incompatibilized NR/EPDM/phlogopite composite without AEAPS exhibited poorer properties, but NR/EPDM/phlogopite compatibilized by AEAPS and SA showed improved properties. Most properties of the compatibilized NR/EPDM/phlogopite composite were similar to those of the NR/EPDM/CB composite, except for the lower abrasion resistance. The NR/EPDM/phlogopite/AEAPS rubber composite may potentially be used in various applications by replacing expensive fillers, such as CB.

**Keywords:** phlogopite; natural rubber (NR); ethylene-propylene-diene monomer rubber (EPDM); phlogopite; mechanical properties; compatibility

Citation: Lee, S.-H.; Park, S.-Y.; Chung, K.-H.; Jang, K.-S. Phlogopite-Reinforced Natural Rubber (NR)/Ethylene-Propylene-Diene Monomer Rubber (EPDM) Composites with Aminosilane Compatibilizer. *Polymers* **2021**, *13*, 2318. https://doi.org/10.3390/polym13142318

Academic Editor: Kwang-Jea Kim

Received: 2 June 2021
Accepted: 8 July 2021
Published: 14 July 2021

**Publisher's Note:** MDPI stays neutral with regard to jurisdictional claims in published maps and institutional affiliations.

**Copyright:** © 2021 by the authors. Licensee MDPI, Basel, Switzerland. This article is an open access article distributed under the terms and conditions of the Creative Commons Attribution (CC BY) license (https://creativecommons.org/licenses/by/4.0/).

## 1. Introduction

Rubbers with high diene contents, such as natural rubber (NR), polybutadiene, nitrile rubber, and styrene-butadiene rubber, exhibit poor outdoor properties [1–4]. In particular, NR is a natural biosynthesis polymeric rubber composed of isoprene with minute organic impurities and water. Among the high diene concentration-containing rubbers, NR has been extensively exploited, either alone or in combination with other materials, in a wide range of applications, such as in automobiles, trains, tires, conveyor belts, hoses, balls, cushions, gloves, and shoes [1,5–7]. NR features good elasticity (resilience), strength, and processing characteristics, and excellent physical and mechanical properties [8–11]. However, its poor heat, UV, oxygen, and ozone resistance caused by the highly unsaturated polymeric backbone hinders some applications requiring weathering resistance [12–14].

Instead of creating a new singular rubber material, an alternative facile avenue of developing a new advanced material with the required properties is the blending of different rubbers [15,16]. Ethylene-propylene-diene monomer (EPDM) rubber is a less unsaturated elastomer and is polymerized using ethylene and propylene, with a small concentration of nonconjugated diene. EPDM features good aging characteristics, good resistance to weathering, and oxidation and chemical resistance [17,18]. The poor environmental (ozone, heat, and UV) resistance of the NR phase was considerably countered by the EPDM phase without sacrificing the unique properties of NR in NR/EPDM rubber blends [19–21]. EPDMs exhibit a balance among chemical, electrical, thermal, and mechanical properties [22]. Thus, NR/EPDM rubber blends have been extensively employed in various applications, particularly in tire-related parts [23]. However, the difference in the olefin content between NR and EPDM causes their respective cure rates to be incompatible, which adversely affects the mechanical properties of the blends [24–26]. The mechanical properties have been enhanced by many approaches such as epoxidizing NR, grafting a vulcanization inhibitor, grafting an accelerator onto EPDM (modified EPDM), two-stage vulcanization, reactive blending, and incorporating compatibilizers or reinforcing fillers [27–31].

In terms of reinforcing fillers, carbon black (CB), silica, clay, and $CaCO_3$ are widely utilized in rubber systems, similar to thermoplastic polymer systems [32–35]. In particular, NR/EPDM blend systems inevitably require fillers to enhance the mechanical properties, improve processability, add colors, and reduce the cost. The incorporation of fillers into rubbers brings about diverse interactions at the rubber–filler interfaces [36,37]. Among the various fillers, CB and silica are the most widely used [20,38]. Despite their strong interactions with rubbers, they are relatively expensive, and the application of CB has been hindered owing to its black color [37,38]. Therefore, potential fillers such as soybean protein [39], organo-montmorillonite [40], bio-based fibers [41,42], and biochar [43] have been investigated as alternatives.

Mica (dioctahedral: muscovite and paragonite; trioctahedral: biotite and phlogopite) is commonly utilized to improve the mechanical properties, dynamic characteristics, wear resistance, and processability of rubber composites [44–47]. Among mica fillers, phlogopite as a filler has been introduced in applications such as adhesives [48], plastic parts [49–51], and cosmetics [52]. However, phlogopite has not been employed in rubber applications. Thus, in this study, the effects of phlogopite and a coupling agent for the filler–rubber interfaces on mechanical properties were explored and compared with those of carbon black and silica.

## 2. Experiment

### 2.1. Materials

Natural rubber (NR, STR 5L) and ethylene-propylene-diene monomer rubber (EPDM, Keltan KEP-960N(F)) were purchased from PAN STAR Co. (Bangkok, Thailand) and Kumho Polychem Co. (Seoul, Korea), respectively. Carbon black (CB; N330, 28–36 nm, Aditya Birla Chemicals Co., Estado indio, India), phlogopite (LKAB Minerals Co. 40–80 μm, Luleå, Sweden), and silica (3M Co. 20–30 μm, St. Paul, MN, USA) were used as reinforcing fillers. Zinc oxide (ZnO), stearic acid (SA), tetramethylthiuram disulfide (TMTD), N-cyclohexyl-2-benzothiazyl sulfonamide (CBS), and sulfur were purchased from Puyang Willing Chemical Co. (Puyang, China). Aminoethylaminopropyltrimethoxysilane (AEAPS, OFS-6020, Dow Chemical Co., Midland, MI, USA), DI water (BNOChem Co., Cheongju, Korea), and acetic acid (BNOChem Co., Cheongju, Korea) were used for improving the compatibility of the NR/EPDM blend and phlogopite. The AEAPS solution (AEAPSS) was composed of AEAPS, DIW, and acetic acid (30:20:50 wt%).

### 2.2. Rubber Compounding

The pre-mixing (mastication) of NR (62.5 wt%) and EPDM (37.5 wt%) was carried out using an open two-roll mill to produce a blend band form. Fillers (10 parts per hundred resin (phr)) were added to the rubber blend, followed by the incorporation of ZnO (5 phr) and stearic acid (2 phr). In the case of phlogopite addition, AEAPSS (0, 2, 5, and 10 phr) was added together with phlogopite. After the mixture was mixed for 15 min, TMTD (1 phr), CBS (1 phr), and sulfur (1 phr) were added to the mixture and mixed for 10 min.

## 2.3. Curing Characteristic

### 2.3.1. Cure Time ($T_{90}$)

Cure characteristics were examined at 170 °C using a rheometer (DRM-100, Daekyung Engineering Co., Ulsan, Korea) to determine the $T_{90}$ (time at 90% cure extent). The mean values were determined on the basis of five measurements for each sample.

### 2.3.2. Mooney Viscosity

The Mooney viscosity was determined at 125 °C for 4 min using a Mooney viscometer (DWV-200C, Daekyung Engineering Co., Ulsan, Korea). The sample was pre-heated at 125 °C for 1 min prior to the measurements.

## 2.4. Mechanical Properties

### 2.4.1. Tensile Properties

Uniaxial tensile deformation was performed using a universal testing machine (UTM; DUT-500CM, Daekyung Engineering Co., Ulsan, Korea). The tests were performed according to ISO 37. The cross-section of the specimen was 6 × 2 mm, and the gauge length was 40 mm. The specimens were elongated at a constant strain rate of 500 mm/min at 22–24 °C. The mean values were determined based on five specimens. The toughness was determined by integration of stress–strain curves.

### 2.4.2. Hardness

The shore A hardness of the rubber blends and composites was measured according to ISO 48 using a hardness tester (306L, Pacific Transducer instruments, Los Angeles, CA, USA). The mean values were determined based on five specimens.

### 2.4.3. Rebound Resilience

The rebound elasticities of the rubber blends and composites were measured using a ball rebound tester according to ISO 4662. The specimens were maintained at 22–24 °C for 2 h prior to the measurements. The round ball fell onto the samples, and the rebounding height was measured. The mean values were determined based on five specimens.

### 2.4.4. Abrasion Resistance

An abrasion resistance test was conducted using an abrasion tester (DRA-150, Daekyung Engineering Co., Ulsan, Korea). A 2.5 × 2.5 cm sample was placed on a cylindrical tester with a diameter of 15 cm. A cylindrical hammer (470 g) was applied to the sample surface to provide uniform contact forces in abrasive paper of 40 grit (XW341, Deerfos Co., Incheon, Korea). The sample was abraded by rotating it 200 times at 40 rpm. The pristine NR/EPDM was used as a reference sample. The abrasion resistance index (ARI) was calculated based on Equation (1).

$$\text{ARI} = \frac{\Delta m_r \, p_t}{\Delta m_t \, p_r} \times 100 \tag{1}$$

where $\Delta m_r$, $p_r$, $\Delta m_t$, and $p_t$ are the mass loss of the reference compound, density of the reference compound, mass loss of the test rubber, and density of the test rubber, respectively.

### 2.4.5. Morphology

The morphologies of the NR/EPDM blend and NR/EPDM/phlogopite composites were observed by scanning electron microscopy (SEM; Apreo, FEI Co., Hillsboro, OR, USA) at an electron beam voltage of 10.0 kV (at the Center of Advanced Materials Analysis, University of Suwon, Hwaseong, Korea). The surface fractured during tensile tests was coated with a 5–10 nm-thick gold layer using a sputter coater prior to the SEM measurements.

### 2.4.6. FTIR-ATR

Fourier transform infrared (FTIR) spectroscopy (Spectrum Two, PerkinElmer Inc., Waltham, MA, USA) with attenuated total reflection (ATR) mode was performed to investigate the AEAPSS treatment. The thickness of the cured sample was 1 mm. The scan number was 16.

## 3. Results and Discussion

Mooney viscosity tests have been widely utilized to measure the viscosity of raw rubber materials prior to vulcanization. The Mooney viscosity of the pristine NR/EPDM blend was the highest, whereas the incorporation of fillers (silica, CB, and phlogopite) into the blends reduced the Mooney viscosity (Figure 1). The low concentration of filler barely influences the Mooney viscosity. In particular, mica-based fillers with a platy architecture commonly decrease the Mooney viscosity [45,46]. The infiltration of AEAPSS 2 phr into the NR/EPDM/phlogopite elastomeric composites led to compatibilizing effects, thereby increasing the Mooney viscosity. However, after the threshold quantity of 2 phr, the Mooney viscosity decreased with the increasing AEAPSS concentration owing to the plasticization of the excess AEAPSS.

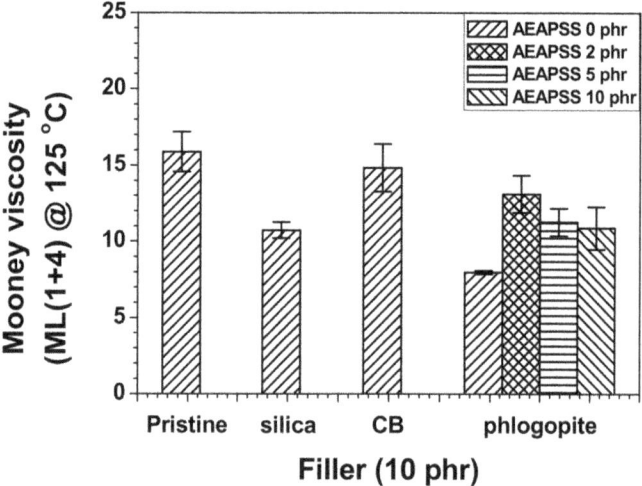

**Figure 1.** Mooney viscosities of the NR/EPDM blends and composites with different fillers and AEAPSS concentrations.

Rubber blends and composites with different formulations exhibited different curing behaviors. The optimum curing time for rubber materials is typically defined as $T_{90}$, which is the time required for the torque to reach 90% of the maximum torque during curing. $T_{90}$ is related to the time required for the development of the optimum properties. The curing behaviors and $T_{90}$ were noticeably influenced by the filler incorporation because of the filler–rubber matrix interactions. The maximum torque for filler-reinforced rubbers typically decreases as a function of temperature. Figure 2 shows the $T_{90}$ values of the NR/EPDM blends and composites with different fillers. The $T_{90}$ values of the NR/EPDM/silica and NR/EPDM/CB composites were lower than those of the pristine NR/EPDM blend, whereas the incorporation of phlogopite into the blend without AEAPSS increased the $T_{90}$ value. The NR/EPDM/phlogopite/AEAPSS composites showed the lowest $T_{90}$ values among the other filler-embedded NR/EPDM composites. This indicates that the amine moieties of AEAPS accelerated the reaction rate of vulcanization [53,54].

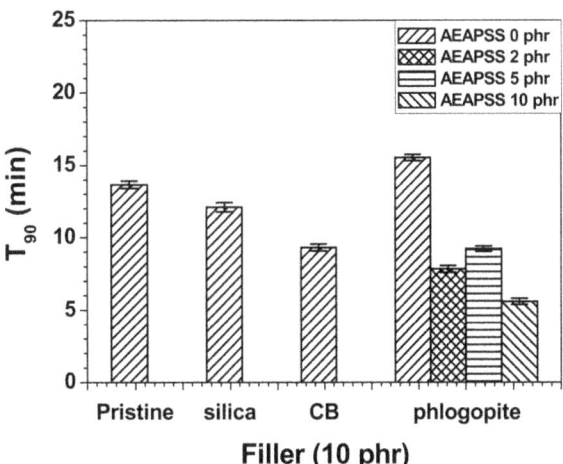

**Figure 2.** $T_{90}$ of the NR/EPDM blends and composites with different fillers and AEAPSS concentrations.

Among various properties, the mechanical properties of rubber composites are crucial, especially for tire and automobile applications. For instance, the tensile strength exhibits the maximum stress before material failure in those applications. Figure 3a shows the tensile strengths of the NR/EPDM blend and composites. The incorporation of silica and CB into the NR/EPDM blend improved the tensile strength, whereas the tensile strength of the NR/EPDM/phlogopite composite barely changed, compared with that of the pristine NR/EPDM blend. This indicates incompatibility between the phlogopite filler and rubbers. The additional infiltration of AEAPSS into the NR/EPDM blend gradually enhanced the tensile strength of the composites as a function of the phlogopite concentration owing to the compatibilizing effect. The elongation at break of each composite except 2 phr AEAPSS was higher than that of the neat NR/EPDM blend, as shown in Figure 3b. The concentration of 2 phr AEAPSS was insufficient to coat the phlogopite. In the absence of SA (red box in Figure 3a–d), the compatibilizing effect was reduced, thereby resulting in a decrease in mechanical properties. The tensile moduli of NR/EPDM/silica and NR/EPDM/CB showed little change, whereas NR/EPDM/phlogopite, even without AEAPSS, portrayed a slight enhancement in the tensile modulus, as shown in Figure 3c. Analogous to the tensile strength results, the toughness of the composites was higher than that of the neat blend. The compatibilized NR/EPDM/phlogopite/AEAPSS 10 phr composite exhibited a higher toughness than the NR/EPDM/CB composite (Figure 3d). The stress–strain curves are shown in Figure S1.

The hardness of rubbers typically indicates resistance to localized plastic deformation that is triggered by mechanical indentation (or abrasion). The hardness is determined by a combination of several factors, such as the elastic stiffness, strength, elongation, toughness, ductility, and viscoelasticity. The shore A hardness is routinely utilized for testing rubber materials. Each of the composites containing each filler exhibited greater hardness than the NR/EPDM blend, as shown in Figure 4. The incorporation of AEAPSS into the NR/EPDM/phlogopite composite slightly increased its hardness. The abrasion resistance indexes (ARIs) of the blends and composites are shown in Figure 5. The ARI of the NR/EPDM/CB composite was lower than that of the neat blend, whereas the incorporation of silica and phlogopite into the blend enhanced the ARIs. Among the various properties, the NR/EPDM/phlogopite composites showed the highest values of the ARI, compared with other composites containing silica or CB.

The elasticity and flexibility of the rubber polymer chains can be confirmed by the rebound resilience tests. The rebound resilience is defined as the ratio of the energy released by the deformation recovery to that required to generate the deformation. It is common for the rebound resilience of rubbers to decrease with increasing filler concentration. The infiltration of fillers into the rubbers reduces the elasticity of the rubber chains, thereby decreasing the resilience properties. Figure 6 shows that the NR/EPDM/silica composite exhibited the lowest reduction in rebound resilience. The effects of CB and phlogopite on the rebound resilience were analogous to each other.

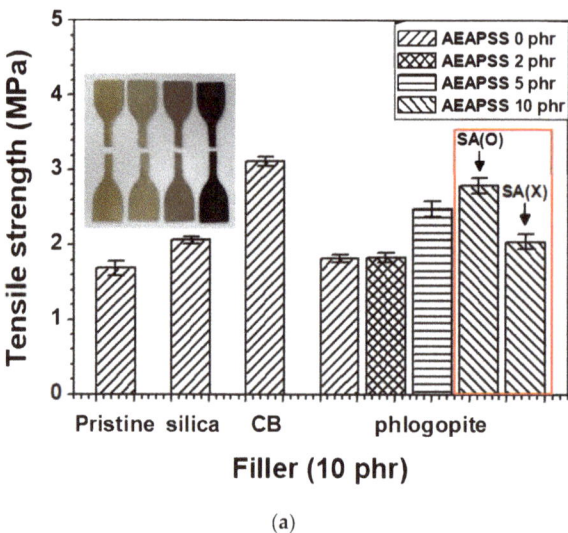

(a)

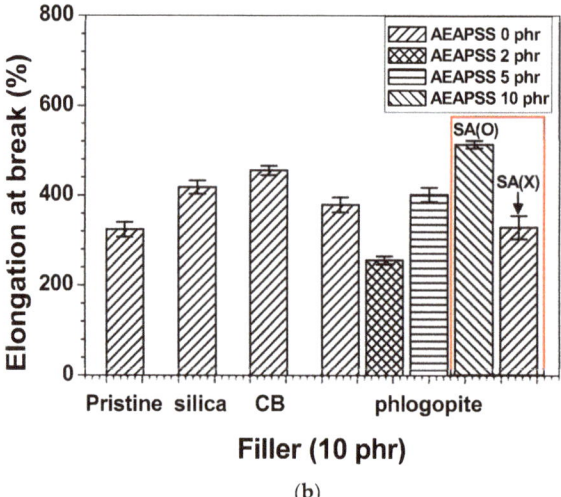

(b)

Figure 3. Cont.

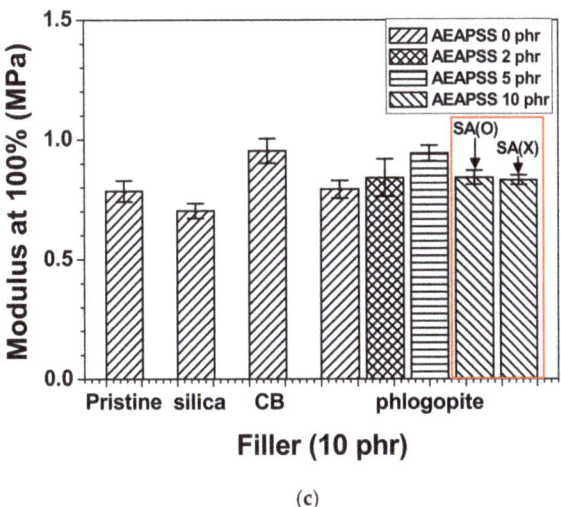

(c)

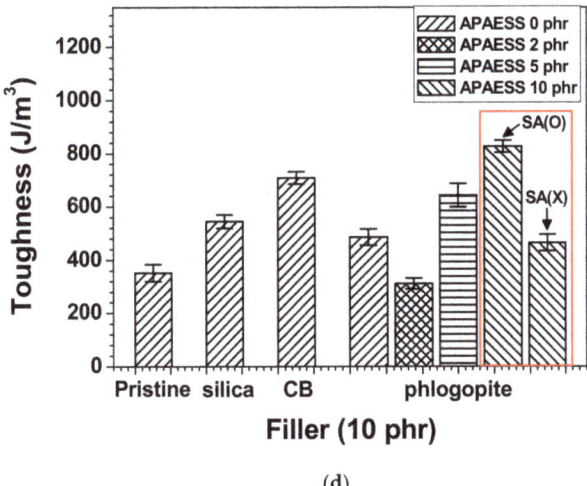

(d)

**Figure 3.** Mechanical properties of the NR/EPDM blends and composites with different fillers and AEAPSS concentrations: (**a**) tensile strength; (**b**) elongation at break; (**c**) tensile modulus at 100%; and (**d**) toughness. The inset of Figure 3a indicates the pristine NR/EPDM, silica-, phlogopite-, and CB-embedded NR/EPDM composites, from left to right.

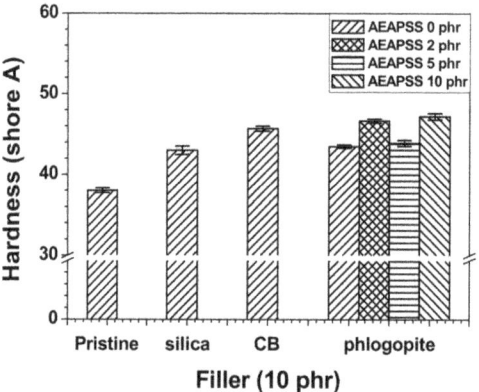

**Figure 4.** Hardness of the NR/EPDM blends and composites with different fillers and AEAPSS concentrations.

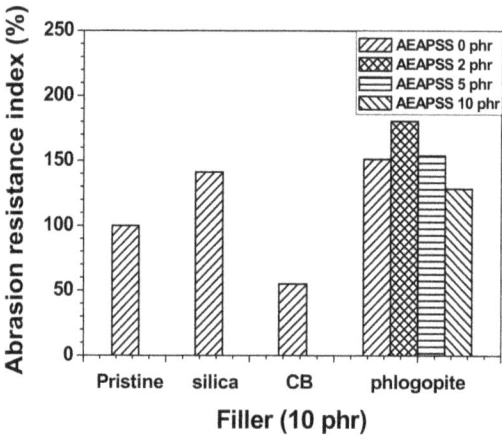

**Figure 5.** ARI of the NR/EPDM blends and composites with different fillers and AEAPSS concentrations.

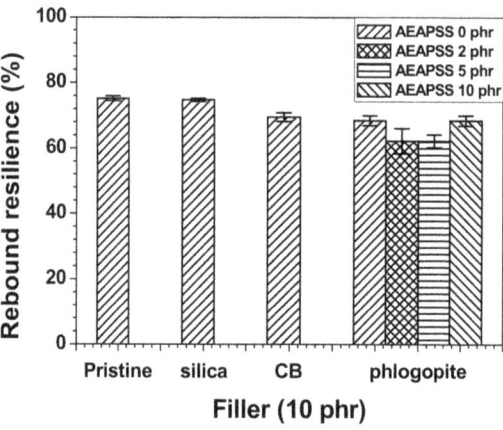

**Figure 6.** Rebound resilience of the NR/EPDM blends and composites with different fillers and AEAPSS concentrations.

Morphological studies of rubber composites are routinely performed by SEM. The morphologies of pure fillers (CB, silica, and phlogopite) and the dispersity of fillers in NR/EPDM systems are shown in Figures S2 and S3, respectively. Figure 7 shows the fractured surfaces of the NR/EPDM blend and composites. The NR/EPDM blend showed a smooth surface with little phase separation, as shown in Figure 7a. The silica- and CB-embedded rubber composites showed good dispersion, with slight agglomeration (Figure 7b,c). The incorporation of AEAPSS into the NR/EPDM/phlogopite composites compatibilized the filler surfaces and rubbers, as observed in Figure 7d,e.

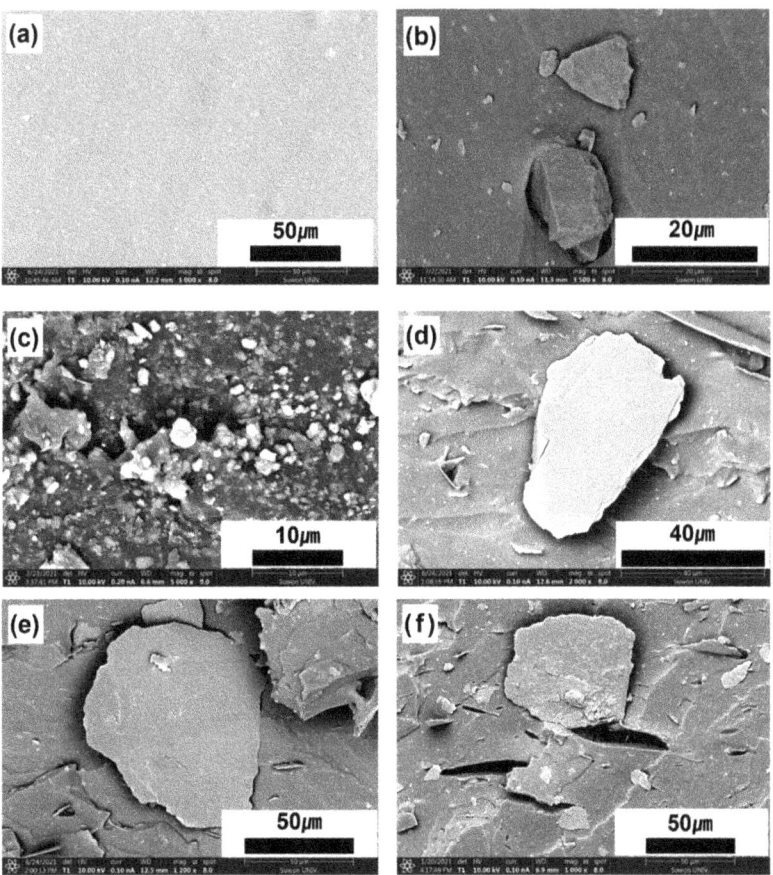

**Figure 7.** SEM images of the NR/EPDM blends and composites with different fillers and AEAPSS concentrations: (**a**) none, (**b**) silica 10 phr, (**c**) CB 10 phr, (**d**) phlogopite 10 phr, (**e**) phlogopite 10 phr/AEAPSS 2 phr, and (**f**) phlogopite 10 phr/AEAPSS 10 phr.

FTIR-ATR was utilized to investigate the effect of AEAPSS on the phlogopite surface treatments, as shown in Figure 8. The broad peaks between 450 and 520 cm$^{-1}$ indicate Si–O and Mg–O for phlogopite. The peaks at 950–990 cm$^{-1}$ and 690 cm$^{-1}$ are ascribed to Si–O for phlogopite [55]. The peak at 1560 cm$^{-1}$ contributing to amine for AEAPS that appeared as 10 phr AEAPSS was added [56,57]. In addition, a peak at 610 cm$^{-1}$ that is associated with the bending vibration of Si–O–Si was observed, probably due to the formation of Si–O–Si between the phlogopite and silane of AEAPSS [58]. On the basis of these results, SA acted as a coupling agent between the rubbers and AEAPSS, whereas AEAPSS acted as a compatibilizing agent between phlogopite and SA, as shown in Figure 9. This interplay created a useful NR/EPDM/phlogopite composite with enhanced properties.

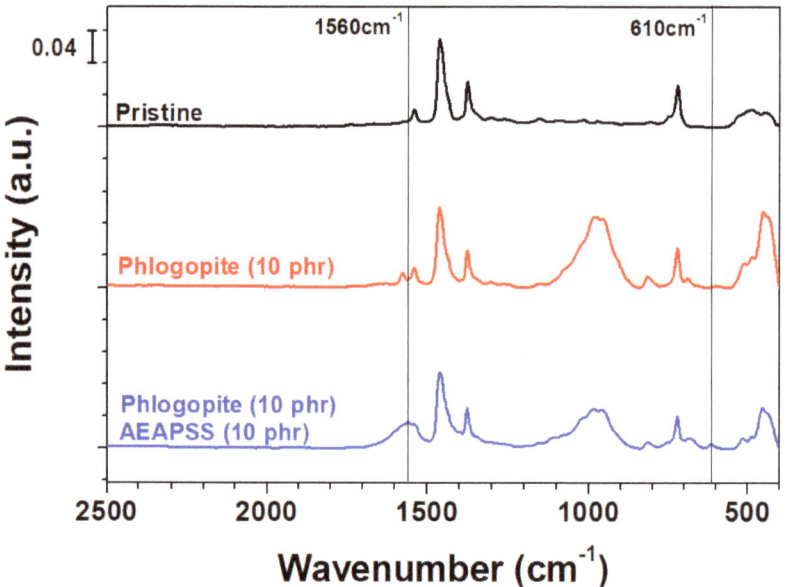

Figure 8. ATR-FTIR spectra of pristine rubber blend and composites consisting of phlogopite 10 phr and phlogopite 10 phr/AEAPS 10 phr.

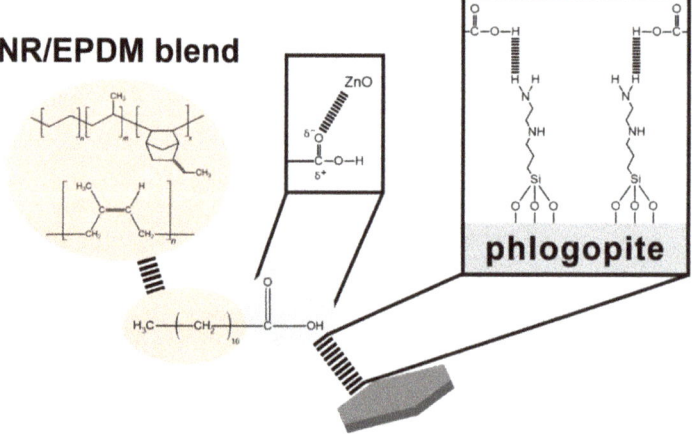

Figure 9. Coupling interactions among phlogopite, rubbers, and stearic acid.

## 4. Conclusions

The effects of phlogopite on various properties (curing behavior, and mechanical, elastic, and morphological properties) of an NR/EPDM blend were examined by comparing silica- and CB-reinforced NR/EPDM composites. In addition, the compatibilizing effect of AEAPSS was investigated for the NR/EPDM/phlogopite composites to further improve the phlogopite-embedded NR/EPDM composite. The combination of SA and AEAPSS provided compatibilizing effects between rubbers and phlogopite. The incompatibilized NR/EPDM/phlogopite composite without AEAPSS showed poorer properties than the NR/EPDM/silica composite, whereas NR/EPDM/phlogopite compatibilized by AEAPSS along with SA was superior to NR/EPDM/silica in terms of most properties. Compared with the NR/EPDM/CB composite, the compatibilized NR/EPDM/phlogopite composite exhibited slightly enhanced or similar properties, except for abrasion resistance. Thus, NR/EPDM/phlogopite/AEAPSS composites may potentially be used in various applications instead of silica- and CB-reinforced NR/EPDM composites.

**Supplementary Materials:** The following are available online at https://www.mdpi.com/article/10.3390/polym13142318/s1, Figure S1: Stress-strain curves of NR/EPDM blends and composites with different fillers and AEAPSS concentrations: (a) Different fillers and (b) different AEAPSS concentrations with and without SA. Figure S2: SEM images of pristine fillers: (a, b) CB, (c) silica, and (d) phlogopite. Figure S3: SEM images of the NR/EPDM composites with different fillers and AEAPSS concentrations: (a) 10 phr CB, (b) 10 phr silica, (c) 10 phr phlogopite, and (d) 10 phr phlogopite/10 phr AEAPSS.

**Author Contributions:** Conceptualization, K.-H.C. and K.-S.J.; methodology, K.-S.J.; validation, S.-H.L. and S.-Y.P.; formal analysis, S.-H.L. and S.-Y.P.; investigation, S.-H.L. and S.-Y.P.; writing—original draft preparation, S.-H.L. and S.-Y.P.; writing—review and editing, K.-S.J.; supervision, K.-H.C. and K.-S.J.; funding acquisition, K.-H.C. and K.-S.J. All authors have read and agreed to the published version of the manuscript.

**Funding:** This work was supported by the Technology Innovation Program (or Industrial Strategic Tech-nology Development Program-Material Components Technology Development Program) (No. 20011433, Extremely cold-resistant anti-vibration elastomer with EPDM) funded by the Ministry of Trade, Industry and Energy (MOTIE, Sejong-si, Korea). This work was supported by the Na-tional Research Foundation of Korea (NRF) grant funded by the Korea government (MSIT) (No. 2021R1G1A1011525, Rapid low-temperature curing of thermoset resins via microwave). This work was supported by a Korea Evaluation Institute of Industrial Technology (KEIT) grant fund-ed by the Korea government (the Ministry of Trade, Industry and Energy, MOTIE) (Anti-fog nano-composite-based head lamp with <10% of low moisture adsorption in surface area; No. 20014475). This work was supported by the Technology Innovation Program (or Industrial Stra-tegic Technology Development Program- Automobile industry technology development) (20015803, High performance composite-based battery pack case for electric vehicles via hybrid structure and weight lightening technology) funded By the Ministry of Trade, Industry & Energy (MOTIE, Korea).

**Institutional Review Board Statement:** Not applicable.

**Informed Consent Statement:** Not applicable.

**Acknowledgments:** This work was supported by the Technology Innovation Program (or Industrial Strategic Technology Development Program-Material Components Technology Development Program) (No. 20011433, Extremely cold-resistant anti-vibration elastomer with EPDM) funded by the Ministry of Trade, Industry and Energy (MOTIE, Sejong-si, Korea). This work was supported by the National Research Foundation of Korea (NRF) grant funded by the Korea government (MSIT) (No. 2021R1G1A1011525, Rapid low-temperature curing of thermoset resins via microwave). This work was supported by a Korea Evaluation Institute of Industrial Technology (KEIT) grant funded by the Korea government (the Ministry of Trade, Industry and Energy, MOTIE) (Anti-fog nano-composite-based head lamp with <10% of low moisture adsorption in surface area; No. 20014475). This work was supported by the Technology Innovation Program (or Industrial Strategic Technology Development Program- Automobile industry technology development) (20015803, High performance composite-based battery pack case for electric vehicles via hybrid structure and weight lightening technology) funded By the Ministry of Trade, Industry & Energy (MOTIE, Korea).

**Conflicts of Interest:** The authors declare no conflict of interest.

# References

1. Rattanasom, N.; Poonsuk, A.; Makmoon, T. Effect of Curing System on the Mechanical Properties and Heat Aging Resistance of Natural Rubber/Tire Tread Reclaimed Rubber Blends. *Polym. Test.* **2005**, *24*, 728–732. [CrossRef]
2. Okamoto, Y. Thermal Aging Study of Carboxyl-Terminated Polybutadiene and Poly(Butadiene-Acrylonitrile)-Reactive Liquid Polymers. *Polym. Eng. Sci.* **1983**, *23*, 222–225. [CrossRef]

3. Liu, X.; Zhao, J.; Yang, R.; Iervolino, R.; Barbera, S. Effect of Lubricating Oil on Thermal Aging of Nitrile Rubber. *Polym. Degrad. Stab.* **2018**, *151*, 136–143. [CrossRef]
4. He, S.; Bai, F.; Liu, S.; Ma, H.; Hu, J.; Chen, L.; Lin, J.; Wei, G.; Du, X. Aging Properties of Styrene-Butadiene Rubber Nanocomposites Filled with Carbon Black and Rectorite. *Polym. Test.* **2017**, *64*, 92–100. [CrossRef]
5. Yip, E.; Cacioli, P. The Manufacture of Gloves from Natural Rubber Latex. *J. Allergy Clin. Immunol.* **2002**, *110*, S3–S14. [CrossRef]
6. Idris, R.; Glasse, M.D.; Latham, R.J.; Linford, R.G.; Schlindwein, W.S. Polymer Electrolytes Based on Modified Natural Rubber for Use in Rechargeable Lithium Batteries. *J. Power Sources* **2001**, *94*, 206–211. [CrossRef]
7. Molnar, W.; Varga, M.; Braun, P.; Adam, K.; Badisch, E. Correlation of Rubber Based Conveyor Belt Properties and Abrasive Wear Rates under 2- and 3-Body Conditions. *Wear* **2014**, *320*, 1–6. [CrossRef]
8. Mott, P.H.; Roland, C.M. Elasticity of Natural Rubber Networks. *Macromolecules* **1996**, *29*, 6941–6945. [CrossRef]
9. Dunuwila, P.; Rodrigo, V.H.L.; Goto, N. Sustainability of Natural Rubber Processing Can Be Improved: A Case Study with Crepe Rubber Manufacturing in Sri Lanka. *Resour. Conserv. Recycl.* **2018**, *133*, 417–427. [CrossRef]
10. Obata, Y.; Kawabata, S.; Kawai, H. Mechanical Properties of Natural Rubber Vulcanizates in Finite Deformation. *J. Polym. Sci. Part 2 Polym. Phys.* **1970**, *8*, 903–919. [CrossRef]
11. Sandberg, O.; Bäckström, G. Thermal Properties of Natural Rubber versus Temperature and Pressure. *J. Appl. Phys.* **1979**, *50*, 4720–4724. [CrossRef]
12. Vinod, V.S.; Varghese, S.; Kuriakose, B. Degradation Behaviour of Natural Rubber–Aluminium Powder Composites: Effect of Heat, Ozone and High Energy Radiation. *Polym. Degrad. Stab.* **2002**, *75*, 405–412. [CrossRef]
13. Zheng, T.; Zheng, X.; Zhan, S.; Zhou, J.; Liao, S. Study on the Ozone Aging Mechanism of Natural Rubber. *Polym. Degrad. Stab.* **2021**, *186*, 109514. [CrossRef]
14. Seentrakoon, B.; Junhasavasdikul, B.; Chavasiri, W. Enhanced UV-Protection and Antibacterial Properties of Natural Rubber/Rutile-TiO$_2$ Nanocomposites. *Polym. Degrad. Stab.* **2013**, *98*, 566–578. [CrossRef]
15. El-Sabbagh, S.H. Compatibility Study of Natural Rubber and Ethylene–Propylene Diene Rubber Blends. *Polym. Test.* **2003**, *22*, 93–100. [CrossRef]
16. Chang, Y.-W.; Shin, Y.-S.; Chun, H.; Nah, C. Effects of Trans-Polyoctylene Rubber (TOR) on the Properties of NR/EPDM Blends. *J. Appl. Polym. Sci.* **1999**, *73*, 749–756. [CrossRef]
17. Kumar, A.; Dipak, G.; Basu, K. Natural Rubber–Ethylene-Propylene-Diene Rubber Covulcanization: Effect of Reinforcing Fillers. *J. Appl. Polym. Sci.* **2002**, *84*, 1001–1010. [CrossRef]
18. Delor-Jestin, F.; Lacoste, J.; Barrois-Oudin, N.; Cardinet, C.; Lemaire, J. Photo-, Thermal and Natural Ageing of Ethylene–Propylene–Diene Monomer (EPDM) Rubber Used in Automotive Applications. Influence of Carbon Black, Crosslinking and Stabilizing Agents. *Polym. Degrad. Stab.* **2000**, *67*, 469–477. [CrossRef]
19. Zaharescu, T.; Meltzer, V.; Vîlcu, R. DSC Studies on Specific Heat Capacity of Irradiated Ethylene-Propylene Elastomers—II. EPDM. *Polym. Degrad. Stab.* **1998**, *61*, 383–387. [CrossRef]
20. Arayapranee, W.; Rempel, G.L. A Comparative Study of the Cure Characteristics, Processability, Mechanical Properties, Ageing, and Morphology of Rice Husk Ash, Silica and Carbon Black Filled 75: 25 NR/EPDM Blends. *J. Appl. Polym. Sci.* **2008**, *109*, 932–941. [CrossRef]
21. Alipour, A.; Naderi, G.; Bakhshandeh, G.R.; Vali, H.; Shokoohi, S. Elastomer Nanocomposites Based on NR/EPDM/Organoclay: Morphology and Properties. *Int. Polym. Process.* **2011**, *26*, 48–55. [CrossRef]
22. Zaharescu, T.; Meltzer, V.; Vîlcu, R. Thermal Properties of EPDM/NR Blends. *Polym. Degrad. Stab.* **2000**, *70*, 341–345. [CrossRef]
23. Cook, S. Compounding NR/EPDM blends for tyre sidewalls. In *Blends of Natural Rubber: Novel Techniques for Blending with Speciality Polymers*; Tinker, A.J., Jones, K.P., Eds.; Springer Netherlands: Dordrecht, The Netherlands, 1998; pp. 184–208. ISBN 978-94-011-4922-8.
24. Sae-oui, P.; Sirisinha, C.; Thepsuwan, U.; Thapthong, P. Influence of Accelerator Type on Properties of NR/EPDM Blends. *Polym. Test.* **2007**, *26*, 1062–1067. [CrossRef]
25. Lewis, C.; Bunyung, S.; Kiatkamjornwong, S. Rheological Properties and Compatibility of NR/EPDM and NR/Brominated EPDM Blends. *J. Appl. Polym. Sci.* **2003**, *89*, 837–847. [CrossRef]
26. Sirqueira, A.S.; Soares, B.G. The Effect of Mercapto- and Thioacetate-Modified EPDM on the Curing Parameters and Mechanical Properties of Natural Rubber/EPDM Blends. *Eur. Polym. J.* **2003**, *39*, 2283–2290. [CrossRef]
27. Hopper, R.J. Improved Cocure of EPDM-Polydiene Blends by Conversion of EPDM into Macromolecular Cure Retarder. *Rubber Chem. Technol.* **1976**, *49*, 341–352. [CrossRef]
28. Azizli, M.J.; Barghamadi, M.; Rezaeeparto, K.; Mokhtary, M.; Parham, S. Compatibility, Mechanical and Rheological Properties of Hybrid Rubber NR/EPDM-g-MA/EPDM/Graphene Oxide Nanocomposites: Theoretical and Experimental Analyses. *Compos. Commun.* **2020**, *22*, 100442. [CrossRef]
29. Chattopadhyay, S. Compatibility Studies on Solution of Polymer Blends by Viscometric and Phase-Separation Technique. *J. Appl. Polym. Sci.* **2000**, *77*, 880–889. [CrossRef]
30. Suma, N.; Joseph, R.; George, K.E. Improved Mechanical Properties of NR/EPDM and NR/ Butyl Blends by Precuring EPDM and Butyl. *J. Appl. Polym. Sci.* **1993**, *49*, 549–557. [CrossRef]
31. Sahakaro, K.; Naskar, N.; Datta, R.N.; Noordermeer, J.W.M. Blending of NR/BR/EPDM by Reactive Processing for Tire Sidewall Applications. I. Preparation, Cure Characteristics and Mechanical Properties. *J. Appl. Polym. Sci.* **2007**, *103*, 2538–2546. [CrossRef]

32. Jang, K.-S. Probing Toughness of Polycarbonate (PC: Ductile and Brittle)/Polypropylene (PP) Blends and Talc-Triggered PC/PP Brittle Composites with Diverse Impact Fracture Parameters. *J. Appl. Polym. Sci.* **2019**, *136*, 47110. [CrossRef]
33. Jang, K.-S. Mechanics and Rheology of Basalt Fiber-Reinforced Polycarbonate Composites. *Polymer* **2018**, *147*, 133–141. [CrossRef]
34. Jang, K.-S. Mineral Filler Effect on the Mechanics and Flame Retardancy of Polycarbonate Composites: Talc and Kaolin. *E-Polymers* **2016**, *16*, 379–386. [CrossRef]
35. Ram, R.; Soni, V.; Khastgir, D. Electrical and Thermal Conductivity of Polyvinylidene Fluoride (PVDF) – Conducting Carbon Black (CCB) Composites: Validation of Various Theoretical Models. *Compos. Part B Eng.* **2020**, *185*, 107748. [CrossRef]
36. Bueche, F. Mullins Effect and Rubber–Filler Interaction. *J. Appl. Polym. Sci.* **1961**, *5*, 271–281. [CrossRef]
37. Fröhlich, J.; Niedermeier, W.; Luginsland, H.-D. The Effect of Filler–Filler and Filler–Elastomer Interaction on Rubber Reinforcement. *Compos. Part Appl. Sci. Manuf.* **2005**, *36*, 449–460. [CrossRef]
38. Ayala, J.A.; Hess, W.M.; Kistler, F.D.; Joyce, G.A. Carbon-Black-Elastomer Interaction. *Rubber Chem. Technol.* **1991**, *64*, 19–39. [CrossRef]
39. Jong, L. Influence of Protein Hydrolysis on the Mechanical Properties of Natural Rubber Composites Reinforced with Soy Protein Particles. *Ind. Crops Prod.* **2015**, *65*, 102–109. [CrossRef]
40. Arroyo, M.; López-Manchado, M.A.; Herrero, B. Organo-Montmorillonite as Substitute of Carbon Black in Natural Rubber Compounds. *Polymer* **2003**, *44*, 2447–2453. [CrossRef]
41. Visakh, P.M.; Thomas, S.; Oksman, K.; Mathew, A.P. Crosslinked Natural Rubber Nanocomposites Reinforced with Cellulose Whiskers Isolated from Bamboo Waste: Processing and Mechanical/Thermal Properties. *Compos. Part Appl. Sci. Manuf.* **2012**, *43*, 735–741. [CrossRef]
42. Pasquini, D.; Teixeira, E.d.M.; Curvelo, A.A.d.S.; Belgacem, M.N.; Dufresne, A. Extraction of Cellulose Whiskers from Cassava Bagasse and Their Applications as Reinforcing Agent in Natural Rubber. *Ind. Crops Prod.* **2010**, *32*, 486–490. [CrossRef]
43. Peterson, S.C. Utilization of Low-Ash Biochar to Partially Replace Carbon Black in Styrene–Butadiene Rubber Composites. *J. Elastomers Plast.* **2013**, *45*, 487–497. [CrossRef]
44. Castro, D.F.; Suarez, J.C.M.; Nunes, R.C.R.; Visconte, L.L.Y. Effect of Mica Addition on the Properties of Natural Rubber and Polybutadiene Rubber Vulcanizates. *J. Appl. Polym. Sci.* **2003**, *90*, 2156–2162. [CrossRef]
45. Escócio, V.A.; Visconte, L.L.Y.; Nunes, R.C.R.; Oliveira, M.G. Rheology and Processability of Natural Rubber Composites with Mica. *Int. J. Polym. Mater. Polym. Biomater.* **2008**, *57*, 374–382. [CrossRef]
46. Roshanaei, H.; Khodkar, F.; Alimardani, M. Contribution of Filler–Filler Interaction and Filler Aspect Ratio in Rubber Reinforcement by Silica and Mica. *Iran. Polym. J.* **2020**, *29*, 901–909. [CrossRef]
47. Ünalan, F.; Gürbüz, Ö.; Nihan, N.; Bilgin, P.; Sermet, B. Effect of Mica as Filler on Wear of Denture Teeth Polymethylmethacrylate (PMMA) Resin. *Balk. J. Stomatol.* **2007**, *11*, 133–137.
48. Verbeek, C.J.R. Highly Filled Polyethylene/Phlogopite Composites. *Mater. Lett.* **2002**, *52*, 453–457. [CrossRef]
49. Cai, L.; Dou, Q. Effect of Filler Treatments on the Crystallization, Mechanical Properties, Morphologies and Heat Resistance of Polypropylene/Phlogopite Composites. *Polym. Compos.* **2019**, *40*, E795–E810. [CrossRef]
50. Verbeek, C.J.R. Effect of Preparation Variables on the Mechanical Properties of Compression-Moulded Phlogopite/LLDPE Composites. *Mater. Lett.* **2002**, *56*, 226–231. [CrossRef]
51. Debnath, S.; De, S.K.; Khastgir, D. Effects of Silane Coupling Agents on Mica-Filled Styrene-Butadiene Rubber. *J. Appl. Polym. Sci.* **1989**, *37*, 1449–1464. [CrossRef]
52. Ohta, S.-I. Synthetic Mica and Its Applications. *Clay Sci.* **2006**, *12*, 119–124. [CrossRef]
53. Magaraphan, R.; Thaijaroen, W.; Lim-ochakun, R. Structure and Properties of Natural Rubber and Modified Montmorillonite Nanocomposites. *Rubber Chem. Technol.* **2003**, *76*, 406–418. [CrossRef]
54. López-Manchado, M.A.; Arroyo, M.; Herrero, B.; Biagiotti, J. Vulcanization Kinetics of Natural Rubber–Organoclay Nanocomposites. *J. Appl. Polym. Sci.* **2003**, *89*, 1–15. [CrossRef]
55. Jenkins, D.M. Empirical Study of the Infrared Lattice Vibrations (1100–350 Cm-1) of Phlogopite. *Phys. Chem. Miner.* **1989**, *16*, 408–414. [CrossRef]
56. Johansson, U.; Holmgren, A.; Forsling, W.; Frost, R.L. Adsorption of Silane Coupling Agents onto Kaolinite Surfaces. *Clay Miner.* **1999**, *34*, 239–246. [CrossRef]
57. Luo, D.; Geng, R.; Wang, W.; Ding, Z.; Qiang, S.; Liang, J.; Li, P.; Zhang, Y.; Fan, Q. Trichoderma Viride Involvement in the Sorption of Pb(II) on Muscovite, Biotite and Phlogopite: Batch and Spectroscopic Studies. *J. Hazard. Mater.* **2021**, *401*, 123249. [CrossRef] [PubMed]
58. Rao Darmakkolla, S.; Tran, H.; Gupta, A.B.; Rananavare, S. A Method to Derivatize Surface Silanol Groups to Si-Alkyl Groups in Carbon-Doped Silicon Oxides. *RSC Adv.* **2016**, *6*, 93219–93230. [CrossRef]

Article

# Anion Exchange Membrane Based on Sulfonated Poly (Styrene-Ethylene-Butylene-Styrene) Copolymers

Hye-Seon Park and Chang-Kook Hong *

Polymer Energy Materials Laboratory, School of Chemical Engineering, Chonnam National University, Gwangju 61186, Korea; parkhseon17@gmail.com
* Correspondence: hongck@chonnam.ac.kr

**Abstract:** Sulfonated poly(styrene-ethylene-butylene-styrene) copolymer (S-SEBS) was prepared as an anion exchange membrane using the casting method. The prepared S-SEBS was further modified with sulfonic acid groups and grafted with maleic anhydride (MA) to improve the ionic conducting properties. The prepared MA-grafted S-SEBS (S-SEBS-g-MA) membranes were characterized by Fourier transform infrared red (FT-IR) spectroscopy and dynamic modulus analysis (DMA). The morphology of the S-SEBS and S-SEBS-g-MA was investigated using atomic force microscopy (AFM) analysis. The modified membranes formed ionic channels by means of association with the sulfonate group and carboxyl group in the SEBS. The electrochemical properties of the modified SEBS membranes, such as water uptake capability, impedance spectroscopy, ionic conductivity, and ionic exchange capacity (IEC), were also measured. The electrochemical analysis revealed that the S-SEBS-g-MA anion exchange membrane showed ionic conductivity of 0.25 S/cm at 100% relative humidity, with 72.5% water uptake capacity. Interestingly, we did not observe any changes in their mechanical and chemical properties, which revealed the robustness of the modified SEBS membrane.

**Keywords:** SEBS; membrane; maleic anhydride; water uptake; impedance spectroscopy; ionic conductivity

## 1. Introduction

The use of renewable energy sources, such as proton exchange membrane fuel cells (PEMFC), is currently growing significantly owing to their environmental, social and economic benefits [1,2]. The conventional renewable energy sources, such as solar and wind, are intermittent and often unpredictable due to their dependency on the weather conditions. These characteristics limit the degree to which utilities can rely upon them, and currently such renewable energy alternatives comprise a small percentage of the primary power sources on the electrical grid [3]. The need to satisfy the energy demand during periods of low energy production has inspired the development of efficient energy storage systems. Recently, the redox flow battery (RFB) has attracted an enormous amount of attention as a promising energy storage system for solar power, nuclear power, emergency uninterruptible power supply (UPS), and for batteries in electric vehicles. In particular, large-scale storage systems are more beneficial as they offer a guaranteed energy supply when using renewable energy sources over a long period. RFBs present several advantages that makes them promising candidates for large-scale energy storage systems; they have energy and power density, capacity which can be designed independently and easily modified even after installation, a moderate operational temperature and a long-life, which makes them highly reliable [3,4].

An RFB consists of electrolyte tanks from which the oxidant and reductant electrolytes are circulated by pumps, through a cell stack comprising a number of connected cells. Each cell comprises an anode and a cathode, which are separated by an ion exchange membrane (IEM). The exchange membrane has been widely used in many fields of life. For example, PEMFC, which includes a membrane, and two electrodes, has grown up with huge attraction because of its simple operation and fuel availability. The PEMFCs have attracted attention from energy devices such as portable, mobile and stationary devices, since it helps in the effective reduction in energy shortage and environmental pollution. The IEM will play an important role in the future of electrical energy generation, which is considered as renewable and clean energy. In addition, the IEM facilitates an improved conductive path between the electrolytes. Therefore, it acts as a membrane separator by preventing cross-mixing and direct chemical reaction of the oxidant and reductant electrolytes from the two reservoir tanks [5,6]. Notably, the ion exchange membrane is a major determinant of redox flow batteries' performance, highlighting the importance of an ideal membrane [7].

Perhaps the most commercially advanced RFB industry that uses a proton-exchange membrane is the vanadium redox-flow battery (VRFB) system [8]. An ion exchange film is an ionic membrane that can selectively separate cations and anions and is an optional transmission film with a functional group that can attract or repel the ions. The IEM has selective permeability to counter-ions due to fixed ions in the actuator. This means that the charged agonist is fixed to the membrane and selectively transmits only the counter-ion with different charges than the agonist, and not the co-ion with the same charge as the agonist. This is called the Donnan exclusion effect—that is, the cation exchange film selectively permeates the cation, and the anion exchange film selectively permeates the anion. The Donnan exclusion effect also causes an unbalanced electrochemical potential difference between the electrolyte and ion exchange membranes, resulting in potential differences in the boundary of the IEMs. This potential difference causes the ions to move until both electrochemical potentials reach equilibrium, which is called the Donnan equilibrium [9]. The most widely used polymer electrolyte membrane is the Nafion 117, a perfluorinated cation exchange film developed by Dupont in 1968. Perfluorometer ion exchange membranes have excellent ion conductivity, chemical stability, and dimensional stability, but crossover occurs due to low ion selection, and above all, they are expensive. To overcome these shortcomings, many studies are being conducted on the development of low-cost hydrocarbon ion exchange membranes [10]. Another important reason for the IEM study is that ion exchange capacity (IEC) will play an important role in the future of electrical energy generation, which is considered as renewable and clean energy [8].

Earlier studies clearly demonstrated that the block copolymers provide excellent separation properties [11–15]. The poly(styrene-ethylene-butylene-styrene) copolymer (SEBS) is one of the promising materials for membrane separators due to its high thermal, chemical, and tunable mechanical properties, and cost effectiveness. The SEBS tri-block polymer has further attracted considerable interest because of its promising proton conducting properties [16,17]. Moreover, due to its simple structure over well-known Nafion 117, the SEBS is considered as a promising anion exchange membrane. It is well known that the key mechanical and electrochemical properties can be tuned via composition of the backbone, hydrocarbon versus fluorocarbons [17,18]. Generally, sulfonated copolymers are synthesized either by a direct copolymerization method or post sulfonated technique [19]. Recently, Mohanty et al. functionalized SEBS membranes through borylation using Suzuki coupling reactions and demonstrated IECs of ca. 2.2 mmol/g [20]. However, the maleic anhydride (MA)-grafted SEBS membrane not yet studied. The MA has ability to enhance the electrochemical properties of SEBS membrane due to its promising solubility properties.

Moreover, while many studies have been conducted on IEM with sulfonation of SESB, the ionic conductivity behavior between sulfonyl and carboxyl functional group in the ionic exchange membrane has not been carried out in previous studies. In this study, we investigated the ionic conductivity behavior of the SEBS membrane with the help of different sulfonyl and carboxylic groups. We synthesized a high-quality sulfonated SEBS-grafted-MA membrane (S-SEBS-g-MA), which permits higher conductivity than conventional Nafion 117, and exhibits a good mechanochemical property with the help of sulfosuccinic acid by cross-linking mechanism. Our electrochemical analysis revealed that the modified membrane shows improved proton conductivity, IEC and water uptake properties. In addition, the dynamic modulus analysis (DMA) result confirms the improved modulus properties.

## 2. Materials and Methods

### 2.1. Materials

The polymeric material used in this experiment was poly(styrene-ethylene-butylene-styrene) copolymer (SEBS, ~118,000 g/mol, 28 wt% polystyrene, Sigma Aldrich, Seoul, Korea). Maleic anhydride (MA, 98.96 g/mol, Daejung, Siheung, Korea) was used as the grafting agent. Chloroform (99.8%, Aldrich, Seoul, Korea) and p-xylene (99%, Sigma Aldrich, Seoul, Korea) were used as a solvent, while dicumyl peroxide (DCP, 98%, Sigma Aldrich, Seoul, Korea), sulfosuccinic acid (SSA, 70 wt%, Sigma Aldrich, Seoul, Korea) and chlorosulfonic acid (99%, Sigma Aldrich, Seoul, Korea) were used as an initiator, the crosslinking agent and sulfonation inhibitor, respectively.

### 2.2. Synthesis of S-SEBS and S-SEBS-g-MA Membrane

#### 2.2.1. Sulfonated SEBS (S-SEBS)

The S-SEBS were prepared by following procedures reported in the previous literature [21,22]. In this process, 10 wt% SEBS was dissolved completely in chloroform by being vigorous stirred for 3 h. The 30 mL of above solution was casted in a circular glass Petri-dish which results in ~150 μm thickness after drying at room temperature for 12 h. The dried film was removed from the glass substrate and cut into a 1 cm × 1 cm square shape. The sulfonation agent was prepared by diluting chlorosulfonic acid in 1,2-dichloroethane and the prepared SEBS membrane was soaking into sulfonation agent for 5 min. Then, the modified membrane sample washed several times with deionized water. The membrane was immersed in deionized water over 24 h, before the tests were carried out.

#### 2.2.2. Synthesis MA Grafted S-SEBS (S-SEBS-g-MA)

We used the backbone-functionalization method for the preparation of S-SEBS-g-MA membrane. The typical scheme used for the preparation is shown in Figure 1. In typical synthesis, 10 wt% SEBS was dissolved in p-xylene for 3 h with continuous stirring. Then, maleic anhydride (MA) of 10 wt% was added into the SEBS solution under nitrogen gas at 135 °C and stirred for 1 h. Next, dicumyl peroxide (DCP) as an initiator in p-xylene was added and then stirred for another 1 h. After cooling at room temperature, 0.5 g sulfosuccinic acid (SSA) is then added in the above solution. The prepared SEBS-g-MA solution was casted on a glass substrate with a thickness of approximately 150 μm and then dried for 12 h at room temperature. The completely dried film was carefully removed from the glass substrate and cut into a square shape (size 1 cm × 1 cm). The SEBS-g-MA membrane was soaked in sulfonation agent for 5 min, and then washing with deionized water several times. The membrane was immersed in deionized water over 24 h before the test was carried out.

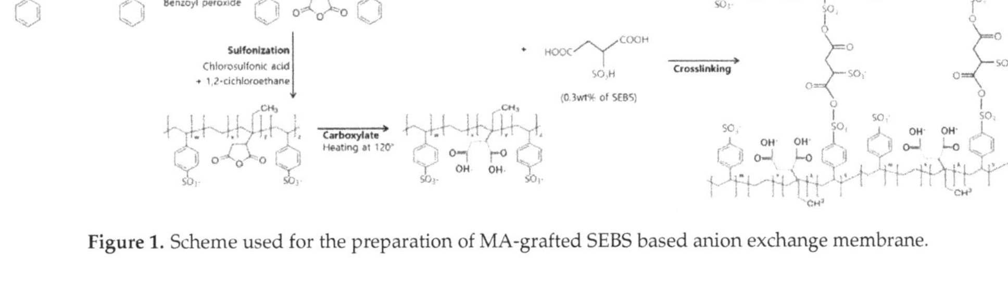

**Figure 1.** Scheme used for the preparation of MA-grafted SEBS based anion exchange membrane.

## 2.3. Characterizations

### 2.3.1. Fourier Transform Infrared (FT-IR) Spectroscopy

The FT-IR spectrometer (Spectrum 400, Perkin Elmer, Gwangju, Korea) was used to investigate the grafting and the functional groups in the synthesized membrane qualitatively.

### 2.3.2. Gel Permeation Chromatography (GPC)

The measurements were conducted using Shodex KF-804, Shodex KF-802, and Shodex KF-801 column (HLC-8320 GPC, Tosoh, Gwangju, Korea) at 40 °C with the eluent flow rate of 1 mL/min and injection of 100 µL. All samples were dissolved in tetrahydrofuran (THF) before measurements. Polystyrene standard was used as a reference.

### 2.3.3. Topographical Analysis

The surface topography of the membranes was investigated using atomic force microscopy (AFM, XE-100, Park system, Gwangju, Korea) operated in tapping mode. For AFM images, the SEBS-based anion exchange membrane films were prepared on the Si wafer at 2000 rpm for 30 s and dried overnight in ambient conditions.

### 2.3.4. Ionic Conductivity

The ionic conductivity of the membrane was obtained by impedance spectroscopy measurement using a conductivity analyzer (Ivuimstat, HS Technologies, Gwangju, Korea) with the frequency range of 1 Hz to 1 MHz at room temperature. Before measurements, the prepared membrane was immersed in deionized water for over 24 h. During measurements, the contact area of the membrane between electrodes was maximized to reduce errors in ionic conductivity measurements. The experimental arrangement for ionic conductivity is shown in Figure 2. For impedance reproducibility and cross-check verification, we successively repeated samples from each set at least five times. The ionic conductivity of the membrane was calculated by using the following equation.

$$\sigma(S/cm) = \frac{L}{RA} \qquad (1)$$

where R is the real impedance taken at zero imaginary impedance in the impedance spectroscopy, and L and A are the thickness and area of the membrane, respectively.

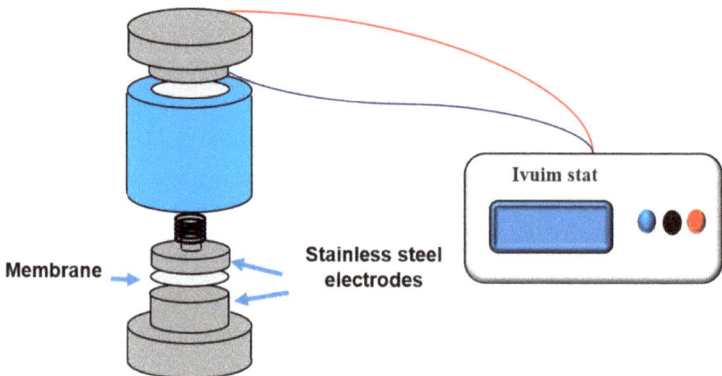

**Figure 2.** Experimental set-up for the ionic conductivity measurements of the different membranes.

### 2.3.5. Ion Exchange Capacity (IEC)

For the ion exchange capacity (IEC) measurement, the dried sample with a certain weight was immersed into 1 M HCl solution overnight and stirred. The ion exchange capacity was determined by titrating the solution with 1 M NaOH solution. The IEC of each sample was measured at least 5 times. The IEC of the membranes was calculated by the following equation.

$$IEC(meq/g) = \frac{M_{O(HCl)} - M_{e\,(HCl)}}{W_{dry}} \quad (2)$$

where $M_{o(HCl)}$ and $M_{e(HCl)}$ are the milliequivalent (meq) of HCl acquired before and after the equilibrium, respectively, and $W_{dry}$ is the mass (g) of the dried membrane.

2.3.6. Sulfonation Degree (SD)

The sulfonation degree (SD) is related to the actual content of sulfonated poly-styrene group of S-SEBS. The SD was calculated using the following equation.

$$SD = \frac{M_P \times IEC}{\{1000 - (M_f \times IEC)\}} \quad (3)$$

where $M_p$ is the molecular weight of the non-functional polymer repeat unit (SEBS) and $M_f$ is the molecular weight of the functional group ($SO_3H$). The values for $M_p$ and $M_f$ are 18,000 and 81, respectively.

2.3.7. Water Uptake

To evaluate the water uptake of the membrane, the sample with a certain weight was immersed in deionized water for 24 h at room temperature. After 24 h, the membrane was taken out and the water on the surface was quickly wiped using filter paper, and then the membrane was weighed again. The water uptake was calculated according to the following equation. The water uptake of each sample was measured at least 5 times and we used the average value.

$$Water\ uptake(\%) = \frac{W_{wet} - W_{dry}}{W_{dry}} \times 100 \quad (4)$$

where $W_{wet}$ and $W_{dry}$ are the weight of the membrane in the wet and dry state, respectively.

2.3.8. Mechanical Properties

Dynamic mechanical properties of the membranes were measured using dynamic mechanical analysis (DMA, universal V3.5B, TA instruments, Seoul, Korea) in tension mode. Samples were heated to 200 °C at frequency of 1 Hz, with a programmed heating rate of 5 °C/min.

## 3. Results and Discussion

### 3.1. Vibrational and GPC Analysis

Figure 3, containing the FT-IR spectra, shows SEBS peaks at 2916 and 2850 $cm^{-1}$, which represents the $CH_3$ and $CH_2$ stretching. There is another peak at 1454 $cm^{-1}$ that correspond to $-CH_2$, and C=C. S-SEBS has two characteristic peaks at 1226 and 1041 $cm^{-1}$ because of the S=O symmetric stretching vibration and the S=O asymmetric stretching vibration of the $SO_3H$ groups, respectively.

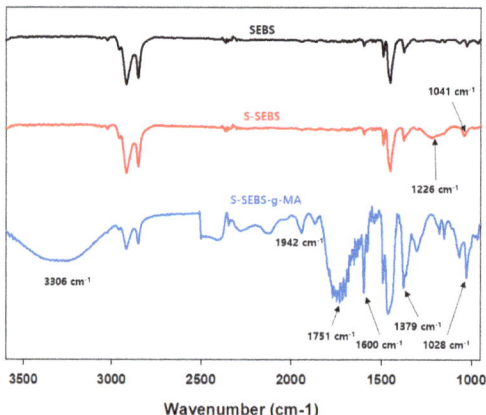

**Figure 3.** The FT-IR spectra of the SBES, S-SEBS, and S-SEBS-g-MA membranes.

However, after grafting by MA, S-SEBS-g-MA has a new peak a at 3306 cm$^{-1}$ because of the hydroxyl group together with some unassociated hydroxyl species in the region 3500–3200 cm$^{-1}$. There are also several other new peaks at 1942 and 1751 cm$^{-1}$ (asymmetric and symmetric carbonyl vibration of MA), 1600 cm$^{-1}$ (C=O stretching of carboxyl group from anhydride), 1379 cm$^{-1}$ (aldehyde), 1028 cm$^{-1}$ (the deformation vibration of the C-H bond) [23,24]. The FT-IR results suggest that the functional groups are responsible for ionic transfer channel of hydroxide ions are successfully grafted on SEBS.

Changes average molecular weight ($M_w$) distribution for SEBS, S-SEBS, SEBS-g-MA, S-SEBS-g-MA, and cross-linked S-SEBS-g-MA (CS-SEBS-g-MA) are presented in Table 1. The $M_w$ of SEBS-g-MA, where MA is grafted onto SEBS, is slightly increased compared to SEBS. However, after sulfonization (S-SEBS, S-SEBS-g-MA), $M_w$ decreased despite the addition of sulfonization compared to the pre-sulfonizaiton samples (SEBS, SEBS-g-MA). From this result, decomposition occurred in the polymer chain during the sulfonization. CS-SEBS-g-MA has the highest molecular weight because the higher molecular weight content of the test specimen increases, which is attributable to the crosslinking of the polymer chain.

**Table 1.** Average of $M_w$ S-SEBS, SEBS-g-MA, S-SEBS-g-MA, and CS-SEBS-g-MA membranes.

| Samples | SEBS | S-SEBS | SEBS-g-MA | S-SEEB-g-MA | CS-SEBS-g-MA |
|---|---|---|---|---|---|
| $M_w$ (g/mol) | 108,589 | 108,493 | 110,179 | 109,986 | 114,241 |

*3.2. Topographical Analysis*

The triblock copolymer SEBS, consisting of hard and soft blocks that usually exhibit a phase-separated morphology, which has been widely studied by AFM techniques [25–29]. It is well known that, for high ionic conductivity of the membrane, a continuous network of a proton conducting phase within the material is essential. To check the surface morphology of membranes, the AFM images of the bare SEBS and modified SEBS membranes were recorded (Figure 4). In Figure 4a, SEBS membrane has a well-defined micro-phase separated morphology, where the dark regions represent the soft polyethylene (PE) block phase. On the other hand, the bright regions represent the stiff polystyrene (PS) phase. Ion exchange membranes of block copolymers consisting of hydrophilic and hydrophobic blocks associate uniformly on microscopic scales and their morphologies are aspherical, cylindrical, or lamellar in shape depending on the relative volume fractions of the constituent components [27,29]. Partially sulfonated SEBS block copolymers are known to demonstrate self-assembling phenomena, which lead to the separation of microphase domains [17,29].

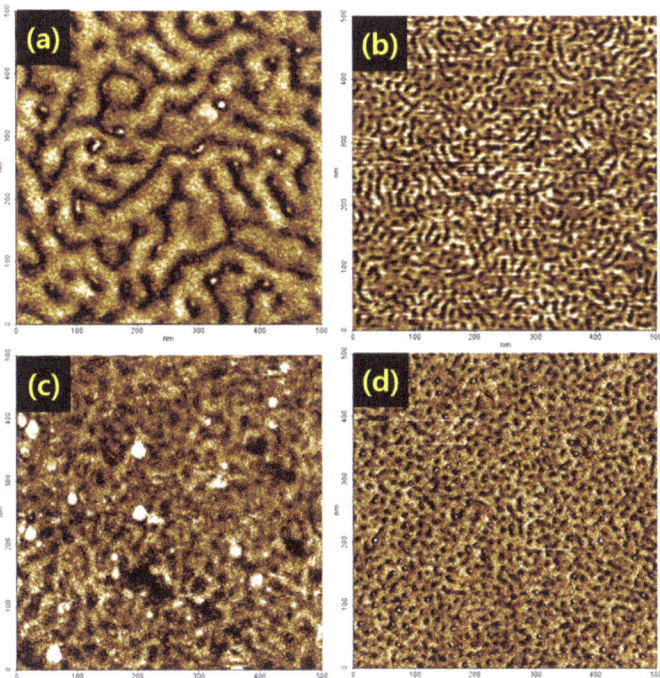

**Figure 4.** The AFM phase images of the modified membranes. (**a**) SEBS, (**b**) SEBS-g-MA, (**c**) S-SEBS, (**d**) S-SEBS-g-MA.

Such microphase separation can result in the formation of continuous ionic channels that enable proton ions transportation through ionic channels. Flexible side chains were also considered to form microphase separation morphology [26]. As a result, we observed well-ordered and more continuously connected nano-channels. Proton transports through these nano-channels with hydrophilic domains lead to good water uptake and ion conductivity. The formation of continuous ionic channels led to proton ion transports through ionic channels. Formation of the continuous ionic channel is dependent on the hydrophilic functional groups such as $SO_3H$ and –COOH. By comparing the SEBS-g-MA membrane with –COOH in Figure 4b and S-SEBS membrane with $SO_3H$ in Figure 4c, different morphologies are observed in the AFM images. These results show that different surface chemical composition characteristics occur depending on the functional groups. Due to this different morphological behavior, ion clusters have different ion transportation speed. As a result, the ionic conductivity of membranes with a –COOH functional group is shown to be different from membranes with a $SO_3H$ functional group. A membrane with two kinds of hydrophobic functional group ($SO_3H$, –COOH) has well-ordered and more continuously connected nano-channels than S-SEBS and SEBS-g-MA, which have single hydrophobic functional groups of $SO_3H$ or COOH, respectively (Figure 4d). These results show that S-SEBS-g-MA is superior to S-SEBS and SEBS-g-MA as an ionic exchange membrane.

*3.3. Ionic Conductivity and Ion Exchange Capacity (IEC)*

The ion exchange membranes contain a high concentration of the fixed ionic groups. On the other hand, the backbone of the membrane is extremely hydrophobic, whereas the charged acid groups are strongly hydrophilic and polar. The hydrophilic domains absorb water and form small clusters distributed throughout the backbone. Proton ions easily permeate the cationic membranes containing fixed negative groups [11]. The ion cluster forms with negative functional groups and the membrane with well-formed ion clusters have high ionic conductivity. Ideal membranes for an RFB should possess good chemical stability and high ionic conductivity. From the impedance results, we observed more hydrophilic functional groups increased, which generates continuous ionic channels responsible for increased ionic conductivity, and decreased resistance, Figure 5.

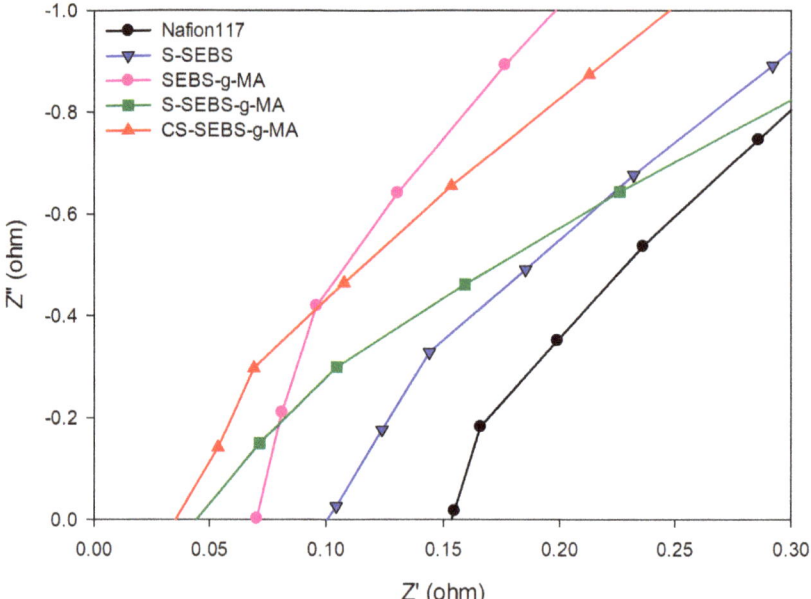

**Figure 5.** Impedance spectra of the Nafion 117, S-SEBS, SEBS-g-MA, S-SEBS-g-MA, and CS-SEBS-g-MA membranes. (Impedance data recorded at least for five sample from each set.).

The resulting values of ionic conductivity and impedance are given in Table 2. As expected, the ionic conductivity is enhanced by increasing the concentration of hydrophilic functional groups. The S-SEBS and S-SEBS-g-MA exhibit improvement in ionic conductivity. The ionic conductivity of the S-SEBS-g-MA increases with sulfonated time and reaches a maximum value 0.18 S/cm at room temperature. As sulfonated time increases, the ionic channels also increase, resulting in an increase in the ionic conductivity. Compared with the ionic conductivity of the S-SEBS membrane (0.1 S/cm), S-SEBS-g-MA shows much higher ionic conductivity.

**Table 2.** Degree of sulfonation, ionic conductivity and IEC values obtained for the bare and modified membranes at room temperature.

| Samples Name | Resistance ($\Omega/cm^2$) | Sulfonation Degree (%) | Ionic Conductivity (S/cm) | IEC (meq) |
|---|---|---|---|---|
| Nafion 117 | 0.145 ± 0.002 | 38.8 ± 0.4 | 0.06 ± 0.001 | 0.9 ± 0.002 |
| S-SEBS | 0.085 ± 0.0001 | 45.6 ± 0.1 | 0.1 ± 0.004 | 2.1 ± 0.004 |
| SEBS-g-MA | 0.06 ± 0.008 | 65.2 ± 0.1 | 0.12 ± 0.004 | 2.8 ± 0.004 |
| S-SEBS-g-MA | 0.047 ± 0.003 | 83.1 ± 0.2 | 0.18 ± 0.004 | 3.36 ± 0.06 |
| CS-SEBS-g-MA | 0.037 ± 0.002 | 102.6 ± 0.2 | 0.25 ± 0.004 | 3.9 ± 0.06 |

This is because S-SEBS has only $SO_3H$, but S-SEBS-g-MA has both $SO_3H$ and COOH, so S-SEBS-g-MA has many more hydrophilic functional groups than S-SEBS. We also used a crosslinking agent containing hydrophilic group to improve both mechanical properties and ion conductivity. As a result, the CS-SEBS-g-MA, which contains more hydrophilic ionic channels than the S-SEBS-g-MA, has an ionic conductivity of 0.25 S/cm. This also indicates that the hydrophilic groups increased the ionic conductivity. Similar to ionic conductivity IEC, the membranes are strongly dependent on the amount of the hydrophilic functional groups. The presence of water plays a vital role in ionic conductivity characteristics; its presence inside the ion exchange membranes offers transport channels for ions. Hence, higher water content will shorten the ion movement pathway and result in higher ionic conductivity. In Table 2 and Figure 6, the IEC value of the S-SEBS membrane is 2.1 meq/g which is much higher than the commercial Nafion 117 membrane, whose IEC is only 0.9 meq/g (Table 2). This means that the IEC also increases as the hydrophilic functional groups increases.

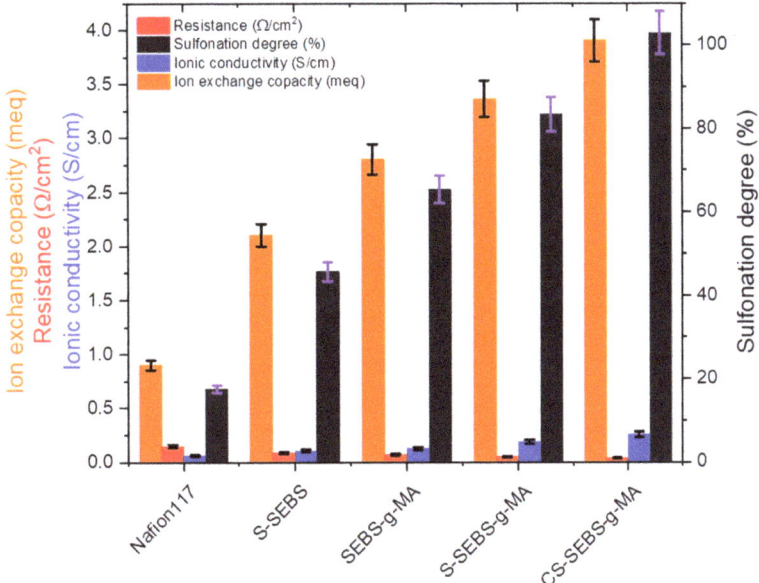

**Figure 6.** Degree of sulfonation, ionic conductivity and IEC of the different membranes studied in this work.

*3.4. Water Uptake*

Next, we investigated the water uptake behavior of all of the above membranes. Figure 7 shows the water uptake of the Nafion 117, S-SEBS, SEBS-g-MA, S-SEBS-g-MA and CS-SEBS-g-MA membrane. Water uptake by the membrane depicts the hydrophobicity of the membrane. The COOH activity in the membrane increases affinity for water more than $SO_3H$, which affected the SEBS-g-MA membrane, causing it to swell more than S-SEBS membrane. The water uptake value for the S-SEBS membrane containing $SO_3H$ functional group is 33% and of the SEBS-g-MA membrane containing COOH functional group is 42% (Figure 7). Thus, S-SEBS exhibits higher ionic conductivity than the SEBS-g-MA membrane, because of ionic conductivity depending on the hydrophilic functional group and the main structure material. Although membranes have identical functional groups, they have differential water uptake values. This is because the water uptake of the membrane is related to the number of hydrophilic groups in the membrane. In other words, the water uptake increases with the increase in the hydrophilic portion of the functional groups in the membrane.

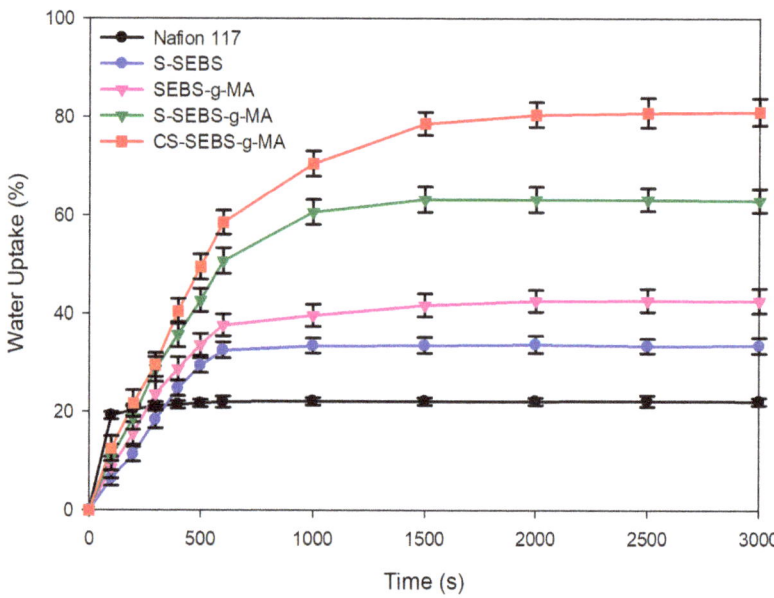

**Figure 7.** Water uptake of the Nafion 117, S-SEBS, SEBS-g-MA, S-SEBS-g-MA, and the CS-SEBS-g-MA membranes as a function of time. Water uptake data were recorded for at least five sample from each sample set.

In comparison with SEBS-g-MA and S-SEBS-g-MA, the S-SEBS-g-MA shows a much higher water uptake than the SEBS-g-MA. The SEBS-g-MA contains only a hydrophilic COOH functional group, while S-SEBS-g-MA comprises two hydrophilic functional groups of $SO_3H$ and COOH. Thus, the higher water uptake of the S-SEBS-g-MA is due to it having more hydrophilic groups. To increase the ionic conductivity of membrane, we used the SSA as a crosslinking agent. SSA has sulfonic acid and carboxylic groups, so SSA play dual roles in the membranes; ion clustering of the hydrophilic group and crosslinking agent. By using the SSA, we increased the level of hydrophilic functionality and the water uptake of CS-SEBS-g-MA to higher than that of S-SEBS-g-MA. Following the results between ionic conductivity and water uptake, we can expect that investigating the functional groups such as sulfonyl and carboxylic acid of the membrane leads to increased ion conductivity. The modified membrane CS-SEBS-g-MA with sulfonyl and carboxylic acid functional groups leads to improved charge density of the modified membrane. From conductivity analysis, we observed significant improvement in the ionic conductivity from 0.1 to 0.18 S/cm, respectively, for the bare S-SEBS to S-SEBS-g-MA membrane with increasing functional groups. Interestingly, this conductivity reached up to 0.25 S/cm for CS-SEBS-g-MA membrane. This clearly indicates that water uptake and ionic conductivity are completely dependent upon the functional groups and cross-linking. This gives rise to the question of how ion conductivity affects the water uptake behavior of the membrane. Increasing the number of functional groups is directly proportional to increasing the ability of $H^+$ transport in the membrane, which leads to improved ionic conductivity and water uptake.

### 3.5. Dynamic Mechanical Analysis (DMA)

A general problem of homogeneous sulfonated styrene main-chain polymers is that these ionomers begin to swell to strong and thus lose their mechanical stability when reaching a certain sulfonated degree. Therefore, reducing the swelling degree of the membranes without lowering their proton conductivity too much is required. These requirements are achieved by cross-linking of the ionomer membrane [30]. Flexible ionomer networks can be built up via ionic crosslinking which contain ionic crosslinks formed by proton transfer. By adding elements with relatively strong electrical voice such as $SOH_3$ and -COOH, it is possible to synthesize colorless and transparent polymer materials by forming an undetermined curved chain structure.

The hydroperoxide sites at the macromolecules can cause radical chain scission [31]. C–H bond of polystyrene is easily attacked by oxygen, forming hydroperoxide radicals [30]. The hydroperoxide sites at the macromolecules can cause radical chain scission. Ionomer crosslinked membranes show reduced brittleness when dried out, compared to uncrosslinked or covalently cross-linked ionomer membranes, which is possibly caused by the flexibleness of ionic cross-links [31]. We used the SSA which contains two types of hydrophilic groups, i.e., $SO_3H$ and COOH as the crosslink agent. The use of the SSA results in a higher ionic conductivity and improved physical properties of the ion exchange membrane. We compared the storage modulus before and after crosslinking of the ion exchange membrane. We observed that the storage modulus of the S-SEBS and S-SEBS-g-MA showed similar behavior, but the storage modulus of the CS-SEBS-g-MA increased by crosslinking between the molecular chains of S-SEBS-g-MA and SSA (Figure 8). The CS-SEBS-g-MA membrane exhibited 50% higher storage modulus than the SEBS-g-MA membrane. These results show that SSA is a good crosslinking agent for the SEBS-g-MA to enhance the mechanical properties and ion conductivity.

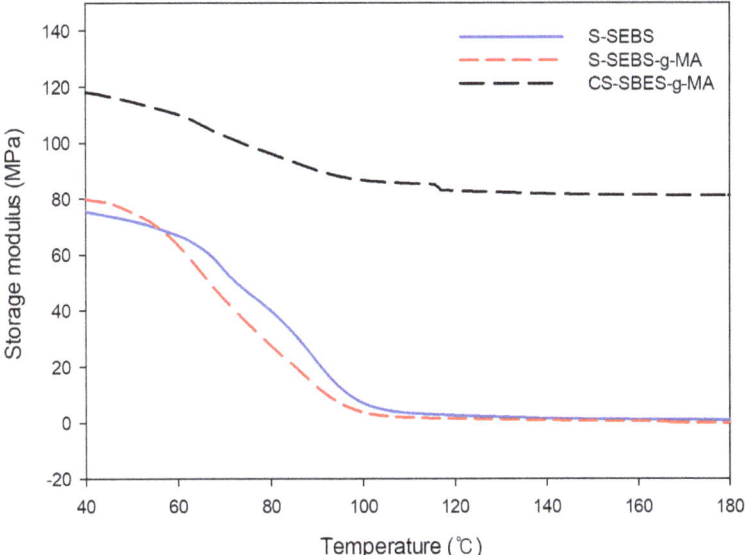

**Figure 8.** Storage modulus of the S-SEBS, S-SEBS-g-MA and CS-SEBS-g-MA.

## 4. Conclusions

In conclusion, we successfully prepared the ionic exchange membrane from SEBS via sulfonation, grafting MA and crosslinking. This study showed that the ionic conductivity properties of the SEBS membrane is improved by chemical modification and cross-linked S-SEBS-g-MA and may be suitable for applying RFB as a membrane in strong acid electrolyte. Sulfonated SEBS block copolymers have continuous ionic channels, while the AFM show direct evidence of well-ordered, nano-sized, and continuous ionic channels. The FT-IR result shows the sulfonic acid groups, carboxylic groups, and grafted MA being successfully introduced into SEBS. Our ion conductivity analysis revealed a significant improvement in the ionic conductivity from 0.1 to 0.25 S/cm, respectively for bare S-SEBS to CS-SEBS-g-MA membrane with increasing functional groups. Additionally, the ionic conductivity, water uptake and IEC of the S-SEBS, SEBS-g-MA, S-SEBS-g-MA and CS-SEBS-g-MA membrane are higher than those of the commercial Nafion 117 membrane. Ionic conductivity of the ionic exchange membrane increases with increasing functional group concentration. These results indicate the modified membrane is a promising candidate for large-scale energy storage systems that can further explored in the future. By manufacturing anion exchange membrane based on non-fluoro polymers such as SEBS, it will be possible to get price competitiveness compared to Nafion 117. Moreover, SEBS-based copolymers can be synthesized by different functional groups, which facilitates the improved ionic conductivity. The lower mechanical properties than conventional Nafion 117 can be improved by blending with alternative polymers or adding crosslink agents. We believe that this method could open up alternative strategy and promising approaches for its further development.

**Author Contributions:** Conceptualization, H.-S.P. and C.-K.H.; methodology, H.-S.P.; validation, H.-S.P. and C.-K.H.; formal analysis, H.-S.P.; investigation, H.-S.P.; data curation, H.-S.P. and C.-K.H.; writing—original draft preparation, H.-S.P.; writing—review and editing, H.-S.P. and C.-K.H.; visualization, H.-S.P.; supervision, C.-K.H.; project administration, C.-K.H.; funding acquisition, C.-K.H. All authors have read and agreed to the published version of the manuscript.

**Funding:** This research was supported by the Jeollanam-do, Korea, under Regional Specialized Industry Development Program of Next-generation stent innovation processing advance technology support project (R&D, B0080621000341).

**Institutional Review Board Statement:** Not applicable.

**Informed Consent Statement:** Not applicable.

**Data Availability Statement:** Not applicable.

**Conflicts of Interest:** Authors have no conflicts to declare.

## References

1. Liang, M.; Liu, Y.; Xiao, B.; Yang, S.; Wang, Z.; Han, H. An analytical model for the transverse permeability of gas diffusion layer with electrical double layer effects in proton exchange membrane fuel cells. *Int. J. Hydrog. Energy* **2018**, *43*, 17880–17888. [CrossRef]
2. Xiao, B.; Wang, W.; Zhang, X.; Long, G.; Chen, H.; Cai, H.; Deng, L. A novel fractal model for relative permeability of gas diffusion layer in proton exchange membrane fuel cell with capillary pressure effect. *Fractals* **2019**, *27*, 1950012. [CrossRef]
3. Parasuramana, A.; Lima, T.M.; Menictas, C.; Skyllas-Kazacos, M. Review of material research and development for vanadium redox flow battery applications. *Electrochim. Acta* **2013**, *101*, 27–40. [CrossRef]
4. Sum, E.; Skylaa-kazacos, M. A study of the V(II)/V(III) redox couple for redox cell applications. *J. Power Sources* **1985**, *15*, 179–190. [CrossRef]
5. Li, X.; Zhang, H.; Mai, Z.; Zhang, H.; Vankelecom, I. Ion exchange membranes for vanadium redox flow battery (VRB) applications. *Energy Environ. Sci.* **2011**, *4*, 1147–1160. [CrossRef]
6. Oei, D.G. Permeation of vanadium cation through anionic and cationic membranes. *J. Appl. Electrochem.* **1985**, *15*, 231–235. [CrossRef]
7. Jia, C.; Liu, J.; Yan, C. A significantly improved membrane for vanadium redox flow battery. *J. Power Sources* **2010**, *13*, 4380–4383. [CrossRef]
8. Yandrasits, M.A.; Lindell, M.J.; Hamroc, S.J. New directions in perfluoroalkyl sulfonic acid–based proton-exchange membranes. *Curr. Opin. Electrochem.* **2019**, *18*, 90–98. [CrossRef]
9. Arora, P.; Zhang, Z. Battery Separators. *Chem. Rev.* **2004**, *104*, 4419–4462. [CrossRef]
10. Bae, B.; Miyatake, K.; Watanabe, M. Watanabe, Synthesis and properties of sulfonated block copolymers having fluorenyl groups for Fuel-Cell applications. *ACS Appl. Mater. Interfaces* **2009**, *1*, 1279–1286. [CrossRef]
11. Li, N.N.; Fane, A.G.; Ho, W.S.W.; Matsuura, T. *Advanced Membrane Technology and Application*; Wiley: Weinheim, Germany, 2011; pp. 841–847.
12. Leibler, L. Theory of microphase separation in block copolymers. *Macromolecules* **1980**, *13*, 1602–1617. [CrossRef]
13. Zeng, Q.H.; Liu, Q.L.; Broadwell, I.; Zhu, A.M.; Xiong, Y.; Tu, X.P. Anion exchange membranes based on quaternized polystyrene-block-poly(ethylene-ran-butylene)-block-polystyrene for direct methanol alkaline fuel cells. *J. Membr. Sci.* **2010**, *349*, 237–243. [CrossRef]

14. Vinodh, R.; Ilakkiya, A.; Elamathi, S.; Sangeetha, D. A novel anion exchange membrane from polystyrene (ethylene butylene) polystyrene: Synthesis and characterization. *Mater. Sci. Eng. B* **2010**, *167*, 43–50. [CrossRef]
15. Castaneda, S.; Ribadeneira, R. Theoretical description of the structural characteristics of the quaternized SEBS Anion-Exchange Membrane Using DFT. *J. Phys. Chem. C* **2015**, *119*, 28235–28246. [CrossRef]
16. Elabd, Y.A.; Napadensky, E.; Sloan, J.M.; Crawford, D.M.; Walker, C.W. Triblock copolymer ionomer membranes: Part1. Methanol and proton transport. *J. Membr. Sci.* **2003**, *217*, 227–242. [CrossRef]
17. Edmondson, C.A.; Fontanella, J.J.; Chung, S.H.; Greenbaum, S.G.; Wnek, G.E. Complex impedance studies of S-SEBS block polymer proton-conducting membrane. *Electrochim. Acta* **2001**, *46*, 1623–1628. [CrossRef]
18. Wintersgill, M.C.; Fontanella, J.J. Complex impedance measurements on Nafion. *Electrochim. Acta* **1998**, *43*, 1533–1538. [CrossRef]
19. Baker, R.W. *Membrane Technology and Applications*, 2nd ed.; Wiley: Weinheim, Germany, 2004; pp. 400–421.
20. Schreiber, M.; Harrer, M.; Whitehead, A.; Bucsich, H.; Dragschitz, M.; Seifert, E.; Tymciw, P. Practical and commercial issues in the design and manufacture of vanadium flow batteries. *J. Power Sources* **2012**, *206*, 483–489. [CrossRef]
21. Mokrini, A.; Acosta, J. Studies of sulfonated block copolymer and its blends. *Polymer* **2001**, *42*, 9–15. [CrossRef]
22. Weiss, R.A.; Sen, A.; Willis, C.L.; Pottick, L.A. Block copolymer ionomers: 1. Synthesis and physical properties of sulphonated poly(styrene-ethylene/butylene-styrene). *Polymer* **1991**, *32*, 1867–1874. [CrossRef]
23. Wootthikanokkhan, J.; Changsuwan, P. Dehydrofluorination of PVDF and Proton Conductivity of the modified PVDF/sulfonated SEBS blend membranes. *J. Met.* **2008**, *18*, 57–62.
24. Won, J.; Choi, S.W.; Kang, Y.S. Structural characterization and surface modification of sulfonated polystyrene–(ethylene–butylene)–styrene triblock proton exchange membranes. *J. Membr. Sci.* **2003**, *214*, 245–257. [CrossRef]
25. Wang, D.; Nakajima, K.; Fujinami, S.; Shibasaki, Y.; Wang, J.Q.; Nishi, T. Characterization of morphology and mechanical properties of block copolymers using atomic force microscopy: Effects of processing conditions. *Polymer* **2012**, *53*, 1960–1965. [CrossRef]
26. Shi, Y.; Zhao, Z.; Liu, W.; Zhang, C. Physically self-cross-linked SEBS anion exchange membranes. *Energy Fuels* **2020**, *34*, 16746–16755. [CrossRef]
27. Laurer, J.H.; Bukovnik, R.; Spontak, R.J. Morphological characteristics of SEBS thermoplastic elastomer gels. *Macromolecules* **1996**, *29*, 5760–5762. [CrossRef]
28. Han, X.; Hu, J.; Liu, H.; Hu, Y. SEBS aggregate patterning at a surface studied by atomic force microscopy. *Langmuir* **2006**, *22*, 3428–3433. [CrossRef] [PubMed]
29. Lu, X.; Steckle, W.P.; Weiss, R.A. Ionic aggregation in a block copolymer ionomer. *Macromolecules* **1993**, *26*, 5876–5884. [CrossRef]
30. Barrera, G.M.; Lopez, H.; Castano, V.M.; Rodriuez, R. Studies on the rubber phase stability in gamma irradiated polystyrene-SBR blends by using FT-IR and Raman spectroscopy. *Radiat. Phys. Chem.* **2004**, *69*, 155–162. [CrossRef]
31. Kerres, J.A. Development of ionomer membranes for fuel cells. *J. Membr. Sci.* **2001**, *185*, 3–27. [CrossRef]

Article

# Preparation of Chemically Modified Lignin-Reinforced PLA Biocomposites and Their 3D Printing Performance

Seo-Hwa Hong, Jin Hwan Park, Oh Young Kim and Seok-Ho Hwang *

Materials Chemistry & Engineering Laboratory, School of Polymer System Engineering, Dankook University, Yongin, Gyeonggi-do 16890, Korea; seohwa8766@naver.com (S.-H.H.); shssh123@gmail.com (J.H.P.); kyobon@dankook.ac.kr (O.Y.K.)
* Correspondence: bach@dankook.ac.kr; Tel.: +82-31-8005-3588

**Abstract:** Using a simple esterification reaction of a hydroxyl group with an anhydride group, pristine lignin was successfully converted to a new lignin (COOH-lignin) modified with a terminal carboxyl group. This chemical modification of pristine lignin was confirmed by the appearance of new absorption bands in the FT-IR spectrum. Then, the pristine lignin and COOH-lignin were successfully incorporated into a poly(lactic acid) (PLA) matrix by a typical melt-mixing process. When applied to the COOH-lignin, interfacial adhesion performance between the lignin filler and PLA matrix was better and stronger than pristine lignin. Based on these results for the COOH-lignin/PLA biocomposites, the cost of printing PLA 3D filaments can be reduced without changing their thermal and mechanical properties. Furthermore, the potential of lignin as a component in PLA biocomposites adequate for 3D printing was demonstrated.

**Keywords:** poly(lactic acid); lignin; maleic anhydride; chemical modification; 3D printing filament

**Citation:** Hong, S.-H.; Park, J.H.; Kim, O.Y.; Hwang, S.-H. Preparation of Chemically Modified Lignin-Reinforced PLA Biocomposites and Their 3D Printing Performance. *Polymers* **2021**, *13*, 667. https://doi.org/10.3390/polym13040667

Academic Editor: Kwang-Jea Kim

Received: 9 February 2021
Accepted: 19 February 2021
Published: 23 February 2021

**Publisher's Note:** MDPI stays neutral with regard to jurisdictional claims in published maps and institutional affiliations.

**Copyright:** © 2021 by the authors. Licensee MDPI, Basel, Switzerland. This article is an open access article distributed under the terms and conditions of the Creative Commons Attribution (CC BY) license (https://creativecommons.org/licenses/by/4.0/).

## 1. Introduction

Additive manufacturing is redefining manufacturing paradigms and has evolved rapidly in recent years. This technology allows for the production of customized parts, prototyping, and material design [1,2]. Although there are several technologies, fused deposition modeling (FDM), one of the most used 3D printing technologies, has attracted significant interest because of its low cost, ease of use, and large availability [3]. The most frequently used commercial materials for 3D printing by FDM are acrylonitrile-butadiene-styrene terpolymer (ABS) and poly(lactic acid) (PLA). Nevertheless, PLA is more widely used, due to issues concerning environmental pollution [4].

PLA is a versatile biopolymer polymerized from a renewable agricultural-based monomer, 2-hydroxy propionic acid (lactic acid), which is synthesized by fermentation of starch-rich materials like sugar beets, sugarcanes, and corn [5,6]. Therefore, this material is a promising and sustainable alternative to petroleum-based synthetic polymers used for a wide range of commodity applications since it can be processed in a similar manner to polyolefins, using the same processing machinery. Nonetheless, the inherent drawbacks of PLA, such brittleness, low heat resistance, high-cost, and slow crystallization, have impeded its widespread use in commercial applications.

To provide a solution to the aforementioned drawbacks, many interesting strategies, such as plasticization [7], block copolymerization [8,9], and compounding with other tougher polymers or elastomers have been utilized [10–18]. Another cost-effective approach to improve the mechanical resistance of PLA at high temperatures is the addition of filler materials to form biocomposites or nanocomposites. Recently, several types of lignin have been incorporated into PLA matrixes by melt-mixing techniques for the development of sustainable biocomposites [19–22]. Evenly applied in the chemically modified lignin, PLA-based biocomposites often show lower mechanical properties, like tensile strength and elongation at break, compared to neat PLA, due to their poor dispersion and adhesion.

Several studies were carried out regarding incorporating lignin in the PLA matrix to form composite filaments for 3D printing, making it evident that lignin's addition often leads to a decrease in tensile strength and elongation [23–30]. Filaments with poor properties will be the death of successful commercialization efforts. Therefore, the development of new lignin-based biocomposite materials for the production of polymeric filaments for FDM is still a challenge.

In this study, we focus on lignin introduced into FDM printing filaments as filler to reduce PLA filament cost. Since, to the best of our knowledge, pristine lignin without any chemical modification cannot aid in improving the physical properties, we selected a simple esterification reaction with maleic anhydride to prepare chemically modified lignin and improve its surface polarity. The carboxy group on the lignin surface is expected to exhibit better adhesion to the PLA matrix, due to its strong hydrogen bonding ability and PLA chain grafting covalently on the lignin surface during processing under high temperatures. Melt blending of PLA with a chemically modified lignin should enable 3D printed filament production. Therefore, the objective of this study is to characterize the thermal, mechanical, and morphological properties of the resulting lignin-based PLA biocomposites, demonstrating their suitability as filament materials for FDM 3D printing.

## 2. Materials and Methods

### 2.1. Materials

PLA pellets (trade name: Ingeo 2003D; $M_w$ (g/mol) = 181,744 [31]; melt flow index (MFI): 6.0 g/10 min; specific gravity: 1.24) was obtained commercially from NatureWorks LLC (Minnetonka, NM, USA). Organosolv lignin (pH = 6.9–7.1; ash <16%) was obtained from BOC Sciences (Shirley, NY, USA). Maleic anhydride and other solvents used in this work were purchased from SAMCHUN Chemicals (Seoul, Korea).

### 2.2. Preparation of Carboxylated Lignin (COOH-Lignin)

First, lignin (10 g) was dispersed in dimethylforamide (DMF, 150 mL) using a homogenizer (T10 basic ULTRA-TURRAX; IKA-Werke GmbH &Co. KG, Staufen, Germany) at 25 °C. Then, maleic anhydride (70 g, 0.71 mol) was added to the solution and stirred vigorously for 6 h at 120 °C. After the crude mixture was cooled to 25 °C, the modified lignin was obtained by centrifugation and washed thoroughly with ethanol and deionized in water several times. Then, the sample was kept in vacuo at 50 °C for a day to dry completely.

### 2.3. 3D Printing

The monofilaments with diameters of about 1.75 mm were extruded using the Wellzoom 3D desktop filament extruder (Shenzhen Mistar Technology Co., Ltd., Shenzhen, China) at 210 °C. A FDM-type 3D printer (model: Ender 3, Shenzhen Creality 3D Technology Co., Ltd., Shenzhen, China) was used to print the prepared filaments into the sample at a printing temperature of 220 °C.

### 2.4. Equipment and Experiments

FT-IR spectra were recorded on a Nicolet iS10 spectrophotometer (Thermo Scientific Co., Ltd., Waltham, MA, USA) over a range of 4000 to 600 cm$^{-1}$. To measure the thermal behavior of samples, differential scanning calorimetry (DSC) analysis was carried out using the DSC Q2000 apparatus (TA Instruments, New Castle, DE, USA). The scan was performed at the heating rate of 20 °C/min under $N_2$ gas atmosphere. Tensile properties were measured according to ASTM D638-10 at 25 °C using an Instron universal testing machine (LSK 100K; Lloyd Instruments Ltd., West Sussex, UK), equipped with a 10 kN load cell at a constant crosshead speed of 5 mm/min. The tensile properties were evaluated from averages of at least five parallel tests. After preparing the specimen sheet by the typical compression molding process, the tensile test specimen was cut by specimen cutting die into a dumbbell shape (ASTM D638, Type V). The fractured surface and profile morphology of the composites were imaged and analyzed using an EM-30 scanning electron microscope (SEM; Coxem, Deajeon, Korea) at 10 kV.

## 3. Results and Discussion

The compatibility of lignin with commercial polymers is generally limited because it has a complicated steric hinderance induced from the crosslinking and disorder of the molecular structure. Thus, chemical modification of lignin can provide an important key strategy to further expand lignin applications [32]. Among the many important functional groups (phenolic hydroxyl, aliphatic hydroxyl, carbonyl, alkyl aryl ether, biphenyl, diaryl ether, phenylpropane, guaiacyl, syringyl, etc.) [33–36] in lignin, the hydroxy group can react with the anhydride group through a well-known esterification reaction (Scheme 1) because the carbon in the anhydride moiety is a strong electrophile, largely due to the fact that the substantial ring strain is relieved when the ring opens upon nucleophilic attack. Under this organic chemistry, maleic anhydride was chosen as the lignin modifier, and we assumed that the carboxy group covalently linked to the lignin can serve as a hydrogen bonding donor and acceptor, as well as a reaction site capable of transesterification with PLA.

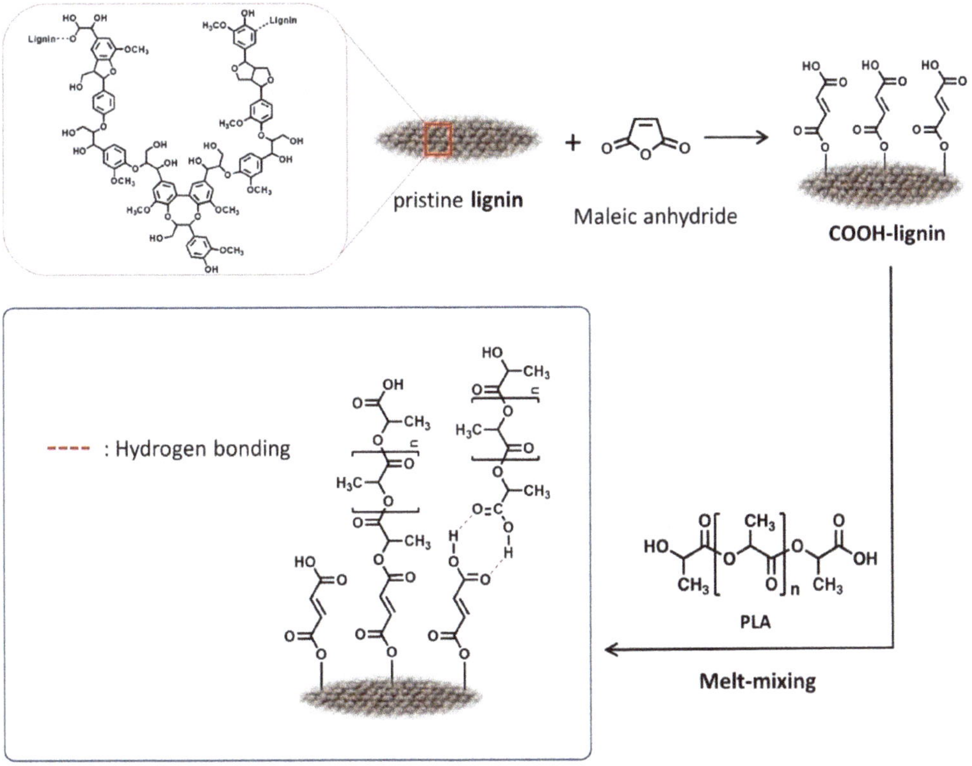

**Scheme 1.** Schematic diagram of the preparation methodology for carboxylated (COOH)-lignin-reinforced poly(lactic acid) (PLA) biocomposites.

To obtain information on chemical changes resulting from the chemical modification of lignin, we used FT-IR spectroscopy. Figure 1 shows the FT-IR absorption spectra based on the molecular vibrations for COOH-lignin and pristine lignin. In the case of pristine lignin, a broad band attributed to aromatic and aliphatic hydroxyl groups in the lignin molecules was detected around 3400–3500 cm$^{-1}$. In addition, the appearance of two bands centered around 2929 and 2848 cm$^{-1}$ was clearly seen, arising from CH$_2$ stretching modes in aromatic methoxy groups and in methyl and methylene groups of side chains. We also found three distinct absorption bands originating from the aromatic skeleton vibration in the lignin molecules around 1586, 1496, and 1451 cm$^{-1}$ [37]. After chemical modification of pristine lignin, the appearance of two new bands around 1722 and 1127 cm$^{-1}$ in the spectrum evidences the formation of ester groups containing one C=O bond and two C–O bonds on the lignin surface. Finally, the vinyl group-related absorption band-like shoulder around 1617 cm$^{-1}$ indicated that maleic anhydride successfully reacted on the surface of the pristine lignin. The FT-IR results indicated that the simple esterification reaction of maleic anhydride on the surface of the lignin particles was successful.

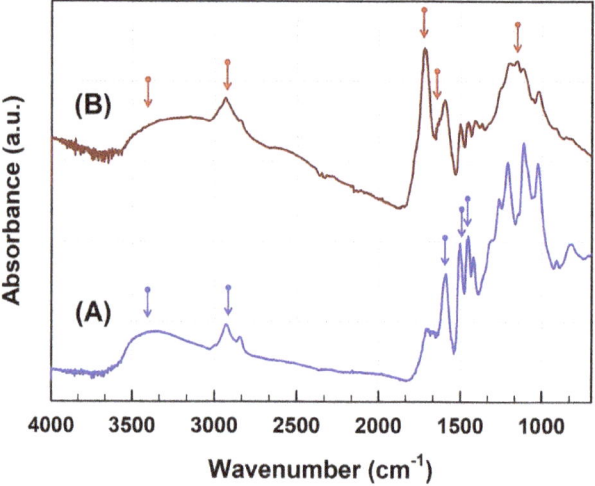

**Figure 1.** FT-IR spectra of pristine lignin (**A**) and COOH-lignin (**B**).

The effect of the presence of pristine lignin and COOH-lignin on the thermal behavior of the PLA matrix was studied using DSC. Table 1 presents thermal characteristics of the melting temperature ($T_m$) observed from the second heating cycle and glass transition temperature ($T_g$) for pristine lignin- or COOH-lignin-reinforced PLA biocomposites. On the basis of Table 1, compatibility between the PLA matrix and lignin domain in the biocomposites should be inferred. The presence of lignin in the PLA matrix did not significantly affect the melting-point depression phenomenon, which decreased from 152.2 to 149.2 °C with increasing content of pristine lignin, consistent with the behavior of lignin as a filler and not a nucleating agent for all the studied PLA-based biocomposites. On the DSC thermograms, the glass transition temperatures of COOH-lignin-reinforced PLA biocomposites showed similar trends to that of pristine lignin-reinforced PLA biocomposites, ranging from 60.3 to 63.4 °C. Further, the sharp $T_g$ transition occurred in all biocomposites. Sometimes, a broad $T_g$ transition in a polymer composite system is caused by the complex molecular structure that leads to multiple relaxation phases within the same transition temperature range [38]. Therefore, even the lignin was chemically modified by maleic anhydride on its surface, though it did not significantly affect the thermal characteristics of the PLA matrix in the biocomposite.

**Table 1.** The melting temperatures ($T_m$) and glass transition temperatures ($T_g$) of poly(lactic acid) (PLA)-based biocomposites with pristine lignin and COOH-lignin content.

| Lignin Content (wt.%) | Pristine Lignin | | COOH-Lignin | |
|---|---|---|---|---|
| | $T_m$ (°C) | $T_g$ (°C) | $T_m$ (°C) | $T_g$ (°C) |
| 5 | 152.2 | 63.4 | 152.1 | 64.4 |
| 10 | 150.1 | 60.6 | 152.5 | 63.9 |
| 15 | 151.7 | 62.5 | 151.5 | 63.4 |
| 20 | 149.2 | 60.3 | 152.2 | 65.0 |
| Std. Dev. | 1.4 | 1.5 | 0.4 | 0.7 |

$T_m$ and $T_g$ of pure PLA measured by DSC in this study are 150.7 and 64.7 °C, respectively.

Dispersibility and interfacial adhesion of filler to the polymer matrix are key factors in determining the final physical properties of a filler-reinforced composite. To evaluate the dispersibility of lignin in the PLA matrix and interfacial adhesion, the morphological characteristics of cryo-fractured pristine lignin- or COOH-lignin-reinforced PLA biocomposites were analyzed using the SEM technique. Figures 2 and 3 show typical SEM micrographs at the fracture surface of the pristine lignin- and COOH-lignin-reinforced PLA biocomposites containing 5, 10, 15, and 20 wt.% of each lignin, respectively. As shown in Figure 2, the fracture surfaces of the pristine lignin/PLA biocomposites were rougher than the COOH-lignin/PLA biocomposites in all ranges of lignin content. In addition, many voids and a wide gap morphology between the lignin domain and the PLA matrix can be easily seen, as evidenced by the observation of debonded areas at the fracture surface, revealing that interfacial adhesion at the interphase does not appear very intense. Somewhat different from the pristine lignin/PLA biocomposites, the COOH-lignin-reinforced PLA biocomposites were smooth and featureless and seemed to show stronger interfacial adhesion with the matrix through the presence of tight adhesive areas in the intermaterial without any room (Figure 3).

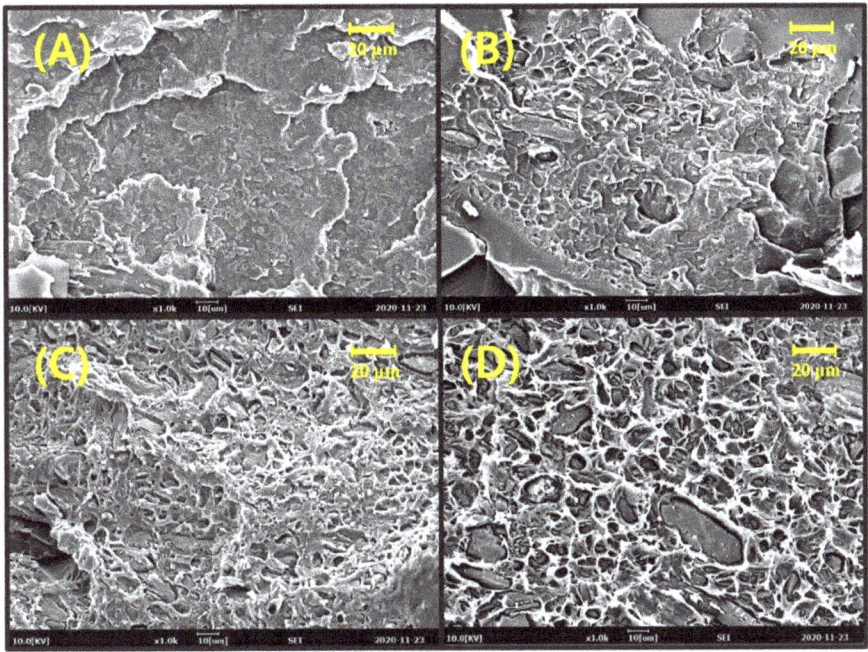

**Figure 2.** SEM micrographs of cryogenically fractured surfaces: poly(lactic acid) (PLA)-based biocomposites with pristine lignin (**A**) 5 wt.%, (**B**) 10 wt.%, (**C**) 15 wt.%, and (**D**) 20 wt.%, respectively.

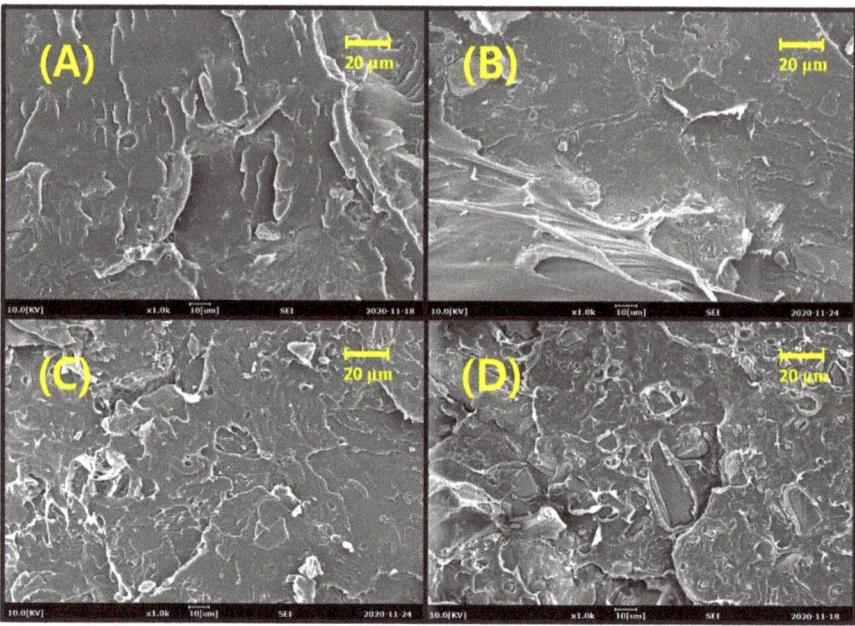

**Figure 3.** SEM micrographs of cryogenically fractured surfaces: poly(lactic acid) (PLA)-based biocomposites with COOH-lignin (**A**) 5 wt.%, (**B**) 10 wt.%, (**C**) 15 wt.%, and (**D**) 20 wt.%, respectively.

To obtain the maximum effect in the filler reinforced composite, the filler should have good dispersibility and interfacial adhesion with the polymer matrix. Here, we investigated the mechanical properties of lignin-reinforced PLA biocomposites following a chemical surface modification of the lignin. Figure 4 presents the dependence of tensile strength and modulus with lignin content for the PLA biocomposites with pristine lignin and COOH-lignin. As shown in Figure 4A, the tensile strength of the pristine lignin/PLA biocomposites decreased linearly with increasing pristine lignin content up to 20 wt.%. This declining trend in tensile strength of the PLA-based biocomposites likely results from poor dispersibility and interfacial adhesion between the pristine lignin domains and PLA matrix [30]. Meanwhile, with COOH-lignin in the PLA matrix, the tensile strength dropped slightly to 50 MPa up to 5 wt.% lignin and then stagnated with further addition of COOH-lignin, manifesting a better COOH-lignin wetting effect with PLA resin. The tensile modulus of the pristine lignin- or COOH-lignin-reinforced PLA biocomposites exhibited similar decreases until 15 wt.% lignin content [Figure 4B]. However, with a 20 wt.% applied to lignin content in the PLA matrix, the tensile modulus showed better values (about 5.0 GPa) compared to both the pristine lignin/PLA biocomposite and pure PLA resin. The above results indicate that COOH-lignin exhibits better dispersibility in the PLA matrix than pristine lignin.

3D printed objects fabricated using COOH-lignin-reinforced PLA biocompoite filaments through a single-screw filament extruder were used to demonstrate the 3D printing behavior and print quality. It was difficult to make the pristine lignin-reinforced PLA biocomposite filaments having a uniform diameter but, all COOH-lignin-reinforced PLA biocompoite filaments exhibited acceptable diameter tolerances (1.41 ± 0.072 mm) for the 3D printer used in this study (Figure 5). The filaments containing COOH-lignin were successfully used to prepare the objects, as illustrated in Figure 6A–E. A closer examination of the surface roughness at smaller scales were obtained from SEM images of the object surfaces [Figure 6a–e]. The surface of the filaments on the 3D printed objects became discernibly rougher and darker in color with increasing COOH-lignin content. This is due to the relatively lower melt strength of COOH-lignin-reinforced PLA biocomposites and the dark brown color of the lignin. However, the surface roughness of each line on the side of the 3D printed object fabricated from PLA biocomposite filament containing more than 15 wt.% COOH-lignin was too rough to bind layer-by-layer, showing a wide gap. This behavior is due to the decreasing melt flow from the nozzle and resultant inadequate adhesion between layers.

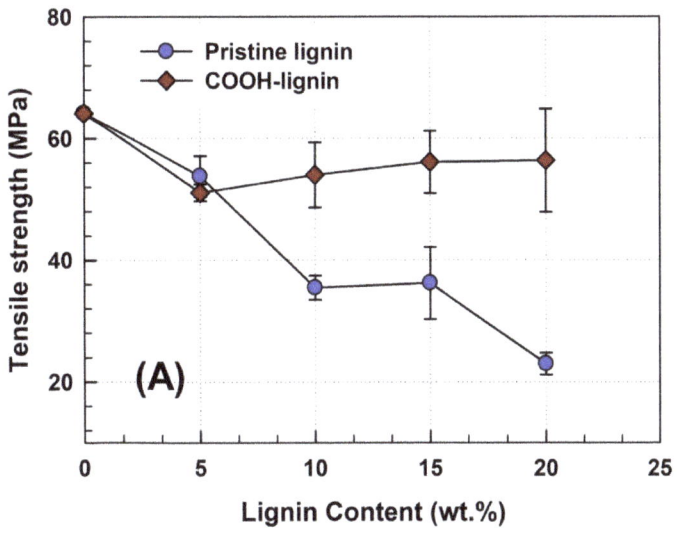

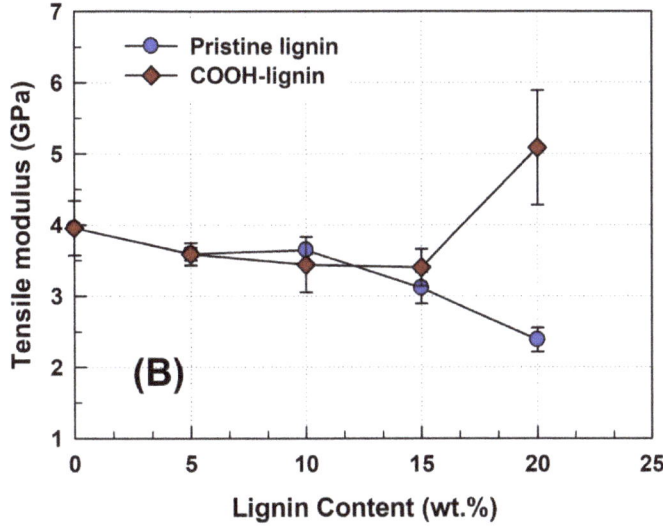

**Figure 4.** Effect of pristine lignin and COOH-lignin content on the tensile strength (**A**) and tensile modulus (**B**) of the PLA-based biocomposites.

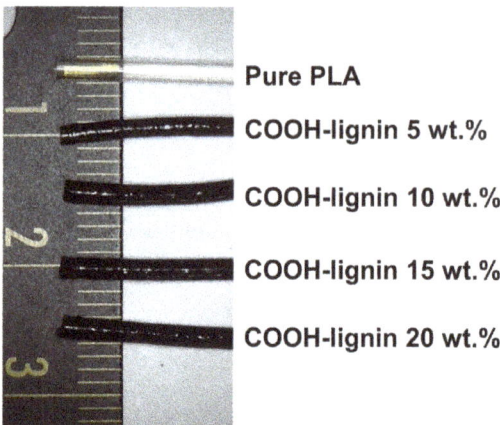

**Figure 5.** Photographs of PLA and COOH-lignin-reinforced PLA biocomposite filaments.

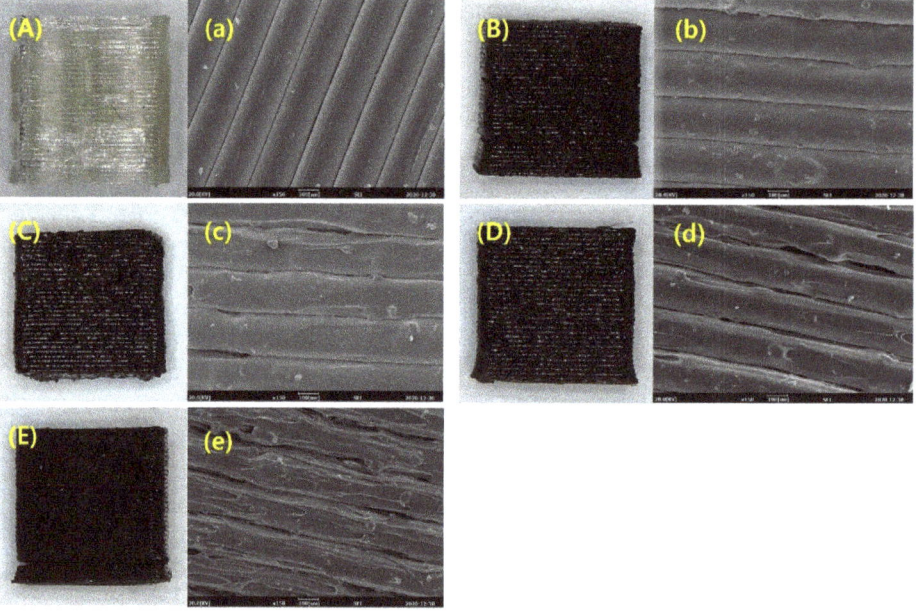

**Figure 6.** Optical image and side-view SEM micrographs of 3D printed objects fabricated by pure PLA (**A**, **a**) and PLA-based biocomposites with COOH-lignin contents of 5 wt.% (**B**, **b**), 10 wt.% (**C**, **c**), 15 wt.% (**D**, **d**), and 20 wt.% (**E**, **e**), respectively.

## 4. Conclusions

By utilizing the reaction of the free hydroxyl group in lignin with the anhydride group, we demonstrated the possibility of realizing surface-modified lignin decorated by carboxyl functional groups. Two different PLA-based biocomposite series with pristine lignin and COOH-lignin were prepared under melt-mixing conditions. The tensile strength of the PLA-based biocomposites indicated better performance of the COOH-lignin/PLA biocomposites than of the pristine lignin/PLA biocomposites, due to improved interfacial adhesion between the COOH-lignin surface and PLA matrix through hydrogen bonding. However, there were no significant differences in the tensile moduli for both PLA-based biocomposites. Interestingly, the tensile modulus of the PLA-based biocomposite containing 20 wt.% COOH-lignin increased suddenly. We also conclude that a COOH-lignin content of 10 wt.% is the most cost effective for 3D FDM filaments. Therefore, PLA-based biocomposites with COOH-lignin can be developed as a new category of polymer materials specially used for the FDM 3D printing method in industrial applications.

**Author Contributions:** Conceptualization, S.-H.H. (S.-H. Hwang); methodology, S.-H.H. (S.-H. Hwang), O.Y.K., S.-H.H. (S.-H. Hong); formal analysis, S.-H.H. (S.-H. Hong), J.H.P.; investigation, S.-H.H. (S.-H. Hong), J.H.P.; data curation, S.-H.H. (S.-H. Hong), J.H.P.; writing–original draft preparation, S.-H.H. (S.-H. Hwang); writing–review and editing, S.-H.H. (S.-H. Hwang), O.Y.K.; supervision, S.-H.H. (S.-H. Hwang); funding acquisition, S.-H.H. (S.-H. Hwang). All authors have read and agreed to the published version of the manuscript.

**Funding:** This research was supported by the GRRC program of Gyeonggi province (GRRC Dankook 2016-B01) and Korea Institute for Advancement of Technology (KIAT) grant funded by the Korea Government (MOTIE) (P0002007, The Competency Development Program for Industry Specialist).

**Institutional Review Board Statement:** Not applicable.

**Informed Consent Statement:** Not applicable.

**Data Availability Statement:** The data presented in this study are available on request from the corresponding author.

**Conflicts of Interest:** The authors declare no conflict of interest.

# References

1. Grisby, W.J.; Scott, S.M.; Plowman-Holmes, M.I.; Middlewood, P.G.; Recabar, K. Combination and processing keratin with lignin as biocomposite materials for additive manufacturing technology. *Acta Biomater.* **2020**, *104*, 95–103. [CrossRef] [PubMed]
2. Nguyen, N.A.; Bowland, C.C.; Naskar, A.K. A general method to improve 3D-printability and inter-layer adhesion in lignin-based composites. *Appl. Mater. Today* **2018**, *12*, 138–152. [CrossRef]
3. McIlroy, C.; Olmsted, P. Disentanglement effects on welding behaviour of polymer melts during the fused-filament-fabrication method for additive manufacturing. *Polymer* **2017**, *123*, 376–391. [CrossRef]
4. Dul, S.; Fambri, L.; Pegoretti, A. Fused deposition modelling with ABS–graphene nanocomposites. *Compos. Part A: Appl. Sci. Manuf.* **2016**, *85*, 181–191. [CrossRef]
5. Jamshidian, M.; Tehrany, E.A.; Imran, M.; Jacquot, M.; Desobry, S. Poly-Lactic Acid: Production, Applications, Nano-composites, and Release Studies. *Compr. Rev. Food Sci.* **2010**, *9*, 552–571. [CrossRef]
6. ElSawy, M.A.; Kim, K.-H.; Park, J.-W.; Deep, A. Hydrolytic degradation of polylactic acid (PLA) and its composites. *Renew. Sustain. Energy Rev.* **2017**, *79*, 1346–1352. [CrossRef]
7. Labrecque, L.V.; Kumar, R.A.; Gross, R.A.; McCarthy, S.P. Citrate esters as plasticizers for poly(lactic acid). *J. Appl. Polym. Sci.* **1997**, *66*, 1507–1513. [CrossRef]
8. Hu, D.S.-G.; Liu, H.-J. Effect of soft segment on degradation kinetics in polyethylene glycol/poly(l-lactide) block copolymers. *Polym. Bull.* **1993**, *30*, 669–676. [CrossRef]
9. Younes, H.; Cohn, D. Phase separation in poly(ethylene glycol)/poly(lactic acid) blends. *Eur. Polym. J.* **1988**, *24*, 765–773. [CrossRef]
10. Liu, G.-C.; He, Y.-S.; Zeng, J.-B.; Xu, Y.; Wang, Y.-Z. In situ formed crosslinked polyurethane toughened polylactide. *Polym. Chem.* **2014**, *5*, 2530–2539. [CrossRef]
11. Ojijo, V.; Ray, S.S.; Sadiku, R. Toughening of Biodegradable Polylactide/Poly(butylene succinate-co-adipate) Blends via in Situ Reactive Compatibilization. *ACS Appl. Mater. Interfaces* **2013**, *5*, 4266–4276. [CrossRef]
12. Jiang, L.; Wolcott, M.P.; Zhang, J. Study of Biodegradable Polylactide/Poly(butylene adipate-co-terephthalate) Blends. *Biomacromolecules* **2006**, *7*, 199–207. [CrossRef]
13. Kang, H.; Qiao, B.; Wang, R.; Wang, Z.; Zhang, L.; Ma, J.; Coates, P.; Coates, P. Employing a novel bioelastomer to toughen polylactide. *Polymer* **2013**, *54*, 2450–2458. [CrossRef]
14. Bitinis, N.; Verdejo, R.; Cassagnau, P.; Lopez-Manchado, M. Structure and properties of polylactide/natural rubber blends. *Mater. Chem. Phys.* **2011**, *129*, 823–831. [CrossRef]
15. Zhang, C.; Wang, W.; Huang, Y.; Pan, Y.; Jiang, L.; Dan, Y.; Luo, Y.; Peng, Z. Thermal, mechanical and rheological properties of polylactide toughened by epoxidized natural rubber. *Mater. Des.* **2013**, *45*, 198–205. [CrossRef]
16. Bhardwaj, R.; Mohanty, A.K. Modification of Brittle Polylactide by Novel Hyperbranched Polymer-Based Nanostructures. *Biomacromolecules* **2007**, *8*, 2476–2484. [CrossRef] [PubMed]

17. Phuong, V.T.; Coltelli, M.-B.; Cinelli, P.; Cifelli, M.; Verstichel, S.; Lazzeri, A. Compatibilization and property enhancement of poly(lactic acid)/polycarbonate blends through triacetin-mediated interchange reactions in the melt. *Polymer* **2014**, *55*, 4498–4513. [CrossRef]
18. Yuryev, Y.; Mohanty, A.K.; Misra, M. Novel biocomposites from biobased PC/PLA blend matrix system for durable applications. *Compos. Part B: Eng.* **2017**, *130*, 158–166. [CrossRef]
19. Gordobil, O.; Egüés, I.; Llano-Ponte, R.; Labidi, J. Physicochemical properties of PLA lignin blends. *Polym. Degrad. Stab.* **2014**, *108*, 330–338. [CrossRef]
20. Thunga, M.; Chen, K.; Grewell, D.; Kessler, M.R. Bio-renewable precursor fibers from lignin/polylactide blends for conversion to carbon fibers. *Carbon* **2014**, *68*, 159–166. [CrossRef]
21. Spiridon, I.; Leluk, K.; Resmerita, A.M.; Darie, R.N. Evaluation of PLA–lignin bioplastics properties before and after accelerated weathering. *Compos. Part B: Eng.* **2015**, *69*, 342–349. [CrossRef]
22. Singla, R.K.; Maiti, S.N.; Ghosh, A.K. Crystallization, Morphological, and Mechanical Response of Poly(Lactic Acid)/Lignin-Based Biodegradable Composites. *Polym. Technol. Eng.* **2015**, *55*, 475–485. [CrossRef]
23. Mimini, V.; Sykacek, E.; Hashim, S.N.A.S.; Holzweber, J.; Hettegger, H.; Fackler, K.; Potthast, A.; Mundigler, N.; Rosenau, T. Compatibility of kraft lignin, organosolv lignin and lignosulfonate with PLA in 3D printing. *J. Wood Chem. Technol.* **2018**, *39*, 14–30. [CrossRef]
24. Gkartzou, E.; Koumoulos, E.P.; Charitidis, C.A. Production and 3D printing processing of bio-based thermoplastic filament. *Manuf. Rev.* **2017**, *4*, 1–14. [CrossRef]
25. Tanase-Opedal, M.; Espinosa, E.; Rodríguez, A.; Chinga-Carrasco, G. Lignin: A Biopolymer from Forestry Biomass for Biocomposites and 3D Printing. *Materials* **2019**, *12*, 3006. [CrossRef]
26. Domínguez-Robles, J.; Martin, N.K.; Fong, M.L.; Stewart, S.A.; Irwin, N.J.; Rial-Hermida, M.I.; Donnelly, R.F.; Larrañeta, E. Antioxidant PLA Composites Containing Lignin for 3D Printing Applications: A Potential Material for Healthcare Applications. *Pharmer* **2019**, *11*, 165. [CrossRef]
27. Wasti, S.; Triggs, E.; Farag, R.; Auad, M.; Adhikari, S.; Bajwa, D.; Li, M.; Ragauskas, A.J. Influence of plasticizers on thermal and mechanical properties of biocomposite filaments made from lignin and polylactic acid for 3D printing. *Compos. Part B: Eng.* **2020**, *205*, 108483. [CrossRef]
28. Yang, J.; An, X.; Liu, L.; Tang, S.; Cao, H.; Xu, Q.; Liu, H. Cellulose, hemicellulose, lignin, and their derivatives as multi-components of bio-based feedstocks for 3D printing. *Carbohydr. Polym.* **2020**, *250*, 116881. [CrossRef]
29. Ji, A.; Zhang, S.; Bhagia, S.; Yoo, C.G.; Ragauskas, A.J. 3D printing of biomass-derived composites: Application and characterization approaches. *RSC Adv.* **2020**, *10*, 21698–21723. [CrossRef]
30. Bhagia, S.; Lowden, R.R.; Erdman, D., III; Rodriguez, M., Jr.; Haga, B.A.; Solano, I.R.M.; Gallego, N.C.; Pu, Y.; Muchero, W.; Kunc, V.; et al. Tensile properties of 3D-printed wood-filled PLA materials using poplar trees. *Appl. Mater. Today* **2020**, *21*, 100832. [CrossRef]
31. de Almeida, J.F.M.; da Silva, A.L.N.; Escócio, V.A.; da Fonseca Thomé, A.H.M.; De Sousa, A.M.F.; Nascimento, C.R.; Bertolino, L.C. Rheological, mechanical and morphological behavior of polylactide/nano-sized calcium carbonate composites. *Polym. Bull.* **2016**, *73*, 3531–3545. [CrossRef]
32. Cicala, G.; Latteri, A.; Saccullo, G.; Recca, G.; Sciortino, L.; Lebioda, S.; Saake, B. Investigation on Structure and Thermomechanical Processing of Biobased Polymer Blends. *J. Polym. Environ.* **2016**, *25*, 750–758. [CrossRef]
33. Yan, L.; Cui, Y.; Gou, G.; Wang, Q.; Jiang, M.; Zhang, S.; Hui, D.; Gou, J.; Zhou, Z. Liquefaction of lignin in hot-compressed water to phenolic feedstock for the synthesis of phenol-formaldehyde resins. *Compos. Part B: Eng.* **2017**, *112*, 8–14. [CrossRef]
34. Erdtman, H. *Lignins: Occurrence, Formation, Structure and Reactions*; Sarkanen, K.V., Ludwig, C.H., Eds.; John Wiley & Sons: New York, NY, USA, 1971; p. 916.
35. Graupner, N.; Fischer, H.; Ziegmann, G.; Müssig, J. Improvement and analysis of fibre/matrix adhesion of regenerated cellulose fibre reinforced PP-, MAPP- and PLA-composites by the use of Eucalyptus globulus lignin. *Compos. Part B: Eng.* **2014**, *66*, 117–125. [CrossRef]
36. Heiss-Blanquet, S.; Zheng, D.; Ferreira, N.L.; Lapierre, C.; Baumberger, S. Effect of pretreatment and enzymatic hydrolysis of wheat straw on cell wall composition, hydrophobicity and cellulase adsorption. *Bioresour. Technol.* **2011**, *102*, 5938–5946. [CrossRef] [PubMed]
37. Lisperguer, J.; Perez, P.; Urizar, S. Structure and thermal properties of lignins: Characterization by infrared spectroscopy and differential scanning calorimetry. *J. Chil. Chem. Soc.* **2009**, *54*, 460–463. [CrossRef]
38. Abdelwahab, M.A.; Taylor, S.; Misra, M.; Mohanty, A.K. Thermo-mechanical characterization of bioblends from polylactide and poly(buthylene adipate-co-terephthalate) and lignin. *Macromol. Mater. Eng.* **2015**, *300*, 299–311. [CrossRef]

Article

# Compression Molding of Thermoplastic Polyurethane Foam Sheets with Beads Expanded by Supercritical CO$_2$ Foaming

Tao Zhang [1], Seung-Jun Lee [2], Yong Hwan Yoo [2], Kyu-Hwan Park [2] and Ho-Jong Kang [1,*]

[1] Department of Polymer Science and Engineering, Dankook University, 152 Jukjeon-ro, Suji-gu, Yongin-si, Gyeonggi-do 16889, Korea; taozhang1214@gmail.com
[2] HDC Hyundai EP R&D Center, 603 Graduate Schools Bldg., Dankook University, 152 Jukjeon-ro, Suji-gu, Yongin-si, Gyeonggi-do 16889, Korea; jjun1984@hdc-hyundaiep.com (S.-J.L.); yooyh@hdc-hyundaiep.com (Y.H.Y.); kyu@hdc-hyundaiep.com (K.-H.P.)
* Correspondence: hjkang@dankook.ac.kr

**Abstract:** Expanded thermoplastic polyurethane (ETPU) beads were prepared by a supercritical CO$_2$ foaming process and compression molded to manufacture foam sheets. The effect of the cell structure of the foamed beads on the properties of the foam sheets was studied. Higher foaming pressure resulted in a greater number of cells and thus, smaller cell size, while increasing the foaming temperature at a fixed pressure lowered the viscosity to result in fewer cells and a larger cell size, increasing the expansion ratio of the ETPU. Although the processing window in which the cell structure of the ETPU beads can be maintained was very limited compared to that of steam chest molding, compression molding of ETPU beads to produce foam sheets was possible by controlling the compression pressure and temperature to obtain sintering of the bead surfaces. Properties of the foam sheets are influenced by the expansion ratio of the beads and the increase in the expansion ratio increased the foam resilience, decreased the hardness, and increased the tensile strength and elongation at break.

**Keywords:** thermoplastic polyurethane; expanded bead; supercritical CO$_2$ foaming; expansion ratio; resilience; hardness

## 1. Introduction

Polymer foams [1,2] are widely used for light weight polymer molded products. Typical processes for making light weight polymer molded products are the Mucell process [3,4] and the bead foam process [5,6]. In the Mucell process, chemical foaming agents [7,8] or physical foaming agents [9–11] are added to the polymer melt and the melt is transferred through the die or into the mold under pressure and cooled in the extrusion or injection molding process. In the bead foam process, expanded beads [12,13] which have already been foamed or expandable beads [14] containing foaming agents which can be foamed are used to prepare foamed products. Incorporation of the foaming agent into the polymer pellet can be carried out by addition in the polymerization process [15,16], by using a high temperature and pressure autoclave to introduce the foaming agent in the supercritical fluid state to polymer pellets [17,18], or by adding the foaming agent to the polymer melt in the extruder and preparing expandable or expanded beads by controlling the cooling condition [19].

Expandable beads are used most widely in the case of polystyrene [20], and expanded beads are used in the case of polypropylene (expanded polypropylene, EPP) [21], polystyrene (expanded polystyrene, EPS) [22], and polyethylene (expanded polyethylene, EPE) [23]. Expanded bead foams are manufactured through a sintering process using foamed polymer beads, which have excellent insulation, heat resistance, impact resistance, and energy absorption. In particular, EPP is widely used for light weight automobile parts due to its mechanical properties, low thermal conduction, and shock absorption properties [24,25]. Recently, interest in expanded thermoplastic polyurethanes (ETPU), which can be used to prepare soft and flexible material and whose properties can easily be controlled in the polymerization process, is increasing [26,27]. These thermoplastic polyurethane foams are excellent flexible materials with high hardness, rebound resilience, excellent mechanical properties, and dynamic shock absorption. In manufacturing polymer foam molded products from expanded beads, especially in the case of EPP, steam chest molding is used [28,29], where high temperature steam is fed into the injection mold to physically sinter and bond the bead surfaces. The critical factor in this process is maintaining the cell structure of EPP while bonding the bead surfaces, thus the temperature of the steam, pressure, and residence time are important variables. Along with the research and development of expanded beads, research on steam chest molding of expanded beads has been reported [30]; the incorporation of hot air along with steam for uniform penetration of the steam to reduce the molding defects from steam variation has also been reported [31]. Steam chest molding is indisputably the best process for molding of expanded beads, but due to high equipment costs, its general applicability is limited and thus, research on diverse methods to fabricate molded foam products appears to be required.

Compression molding was utilized in this study to diversify the methods for fabricating foam products, as it is the most typical and inexpensive fabrication method in polymer processing. Foam sheets were prepared from expanded thermoplastic polyurethane (ETPU) foamed under diverse supercritical $CO_2$ foaming conditions, and the effect of bead foam structure on the characteristics of the foam sheets was studied.

## 2. Materials and Methods

The thermoplastic polyurethane used in this study was Dongsung Corp. (Busan, Korea) aromatic polyether thermoplastic polyurethane (TPU: 6175AP), having a melting point of 150 °C, specific gravity of 1.055 g/cm$^3$, and Shore A hardness of 78. A lab-designed autoclave (CRS, Anyang, Korea) was used for the foaming of TPU to prepare the ETPU beads. The autoclave was charged with 250 g distilled water, 100 g TPU, 6.70 g tricalcium phosphate (TCP, Sigma-Aldrich, Merck KGaA, Darmstadt, Germany) stabilizer, and 0.13 g sodium dodecylbenzenesulfonate (SDBS, Sigma-Aldrich, Merck KGaA, Darmstadt, Germany) dispersing agent, then $CO_2$ was pumped in with a high-pressure pump (CRS, Anyang, Korea). In order to obtain supercritical $CO_2$, the temperature was set at 90, 100, 105, or 110 °C and the pressure was set at 75, 80, or 90 bar; the TPU was kept in the autoclave for 30 min, then the pressure was quickly released to atmospheric pressure by opening a ball valve to prepare expanded TPU (ETPU) beads. To prepare TPU foam sheets, a mold cavity measuring 10 cm × 10 cm × 2.0 mm with a temperature control system was mounted on a compression molding machine (QMESYS, QM900A, Uiwang, Korea). Foam sheets were prepared by keeping 15 g ETPU charged mold at 140–150 °C and 3.5–10.5 MPa for 2–15 min to sinter the bead surfaces then quenching in water at 4 °C. A schematic of the foaming process to prepare the ETPU beads and the foam sheet compression molding process is shown in Figure 1.

The water displacement method used to measure the density of all samples was according to ASTM-D792. The foam structure of the ETPU beads prepared under different temperature and pressure conditions was characterized by measuring the cell diameter (D) and the cell density (N) using micrographs obtained with a scanning electron microscope (Coxem EM-30, Daejeon, Korea). The expansion ratio was determined by measuring the density of the pellet before and after foaming ($\rho_{TPU}$, $\rho_{ETPU}$) using an electronic densitometer (SD-200L, Vaughan, ON, Canada) then calculating the expansion ratio ($\Phi$) using the following equation.

$$\Phi = \rho_{TPU}/\rho_{ETPU} \tag{1}$$

Five ETPU foam sheet samples with sizes of 20 mm × 90 mm × 3 mm were prepared for tensile testing at the speed of 10 mm/min. The mechanical properties of the prepared ETPU foam sheets were evaluated by measuring the tensile strength, modulus, and elongation at break as a function of extension ratio using a tensile tester (Lloyd LR30K, Cleveland,

OH, USA) and measuring the Shore A hardness using a Shore hardness tester (BS-392-A, Guangzhou Amittari Instruments Co., Ltd., Guangzhou, China). The rebound properties of the foam sheets were evaluated by dropping a 5 mm ball weighing 0.486 g from 41 cm height ($H_o$) and measuring the height it rebounded ($H$) with a lab-made rebound tester; the ball rebound ratio was calculated according to the following equation.

$$R = H/H_o \times 100(\%) \qquad (2)$$

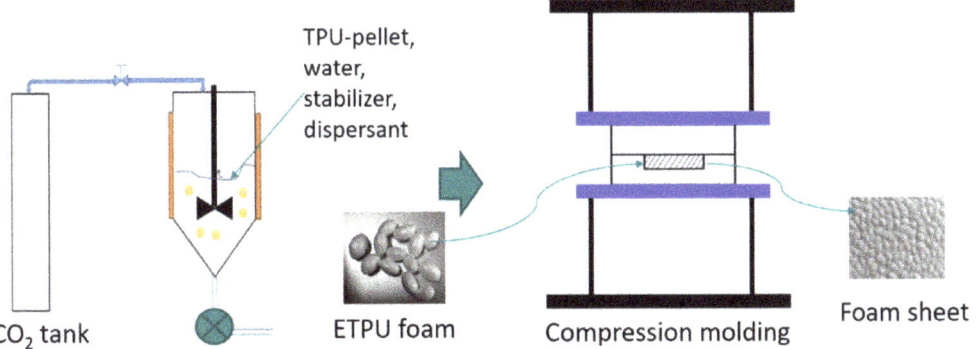

**Figure 1.** Schematic of the $CO_2$ assisted foaming process for preparing expanded thermoplastic polyurethane (ETPU) beads and the compression molding process.

### 3. Results and Discussion

The SEM micrographs of ETPU prepared under different foaming temperatures and pressures are shown in Figure 2. The cell diameter and density measured from Figure 2, and the expansion ratio determined from the density measurements of the pellet before and after foaming shown in Figure 3, reflect the effect of the foaming temperature and pressure on these values. It can be seen in Figure 2 that under the temperature and pressure conditions used in this study, the ETPU foam has a closed cell structure. As can be seen in Figure 2, when the pressure is low (75 bar), the cell is not fully developed and the walls between cells are thick, suggesting that the condition is not adequate for preparing ETPU. The cell size decreases with the rise in pressure and at 90 bar, the cell diameter is 20–60 μm and the cell density is $10^8$ cells/cm$^3$, allowing it to be classified as a fine cell foam [32], regardless of the temperature. In contrast, below 90 bar, the cell diameter is greater than 100 μm and the cell density is $10^6$ cells/cm$^3$, representative of conventional cell foam. This is a result of more nuclei being formed in the TPU at higher pressures, where the same total amount of $CO_2$ is subsequently diffused and the expansion occurring therefrom forms relatively smaller cells. At a fixed pressure, a temperature increase decreases the viscosity of TPU and results in larger cells and lower cell density. The expansion ratio increases with the increase in the pressure and temperature of the foaming process, suggesting that it is more dependent on the cell size compared with cell density. As can be seen in Figure 3c, the expansion ratio of most ETPU obtained in this study is generally below 4, characteristic of high-density foams. However, when the foaming is carried out at 80–90 bar and 110 °C, medium-density foams characterized by expansion ratios of 4–10 are obtained, and when the foaming is carried out at 90 bar and 110 °C, the highest expansion ratio of 7 is obtained. The foam structure, which is dependent on the foaming conditions, will no doubt affect the properties of the foam sheets made from ETPU beads.

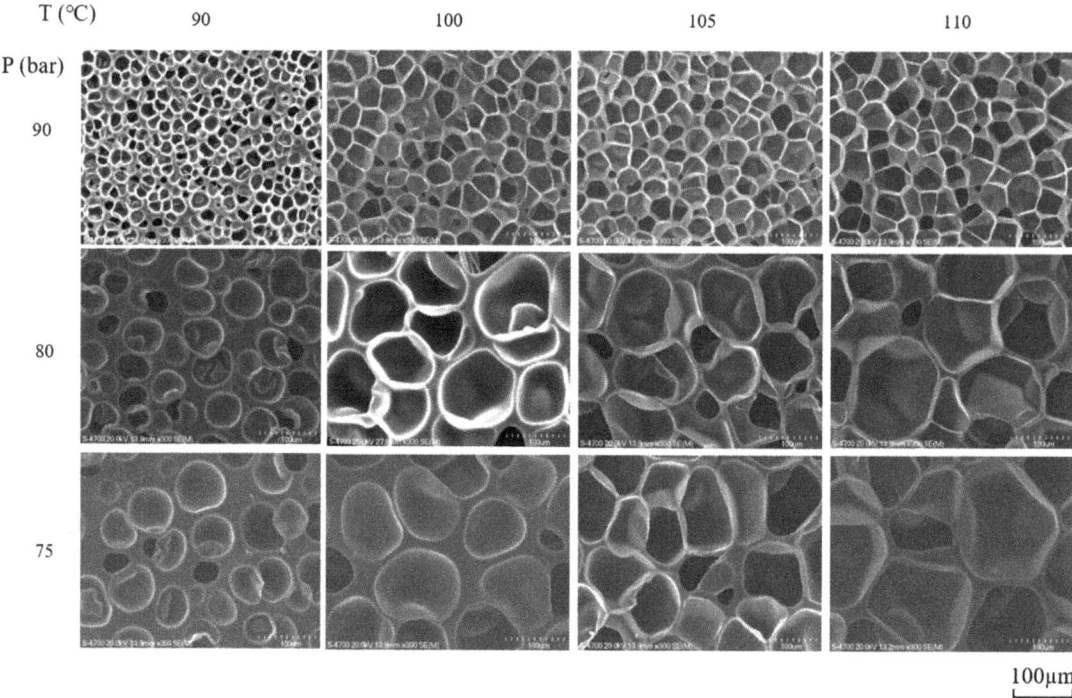

**Figure 2.** SEM micrographs of ETPU prepared at different foaming temperatures and pressures in the supercritical $CO_2$ foaming process.

The effect of the molding temperature on the structure of foam sheets prepared by a 15 min compression molding of ETPU beads at 105 °C and 90 bar can be seen in Figure 4. As can be seen, when compression molded at 140 °C, the fabrication of foam sheets is not possible as sintering does not occur sufficiently, while at 150 °C, melting of the surface of the beads occurs, suggesting that preparation of foam sheets by compression molding should be carried out in a narrow range of temperature slightly below 150 °C, which is the melting point of TPU. The effect of molding time on the formation of the foam sheets at 145 and 150 °C is shown in Figure 5. At 145 °C, sintering of the beads does not occur in 5 min as in the case of molding at 140 °C, but occurs sufficiently in 8–15 min without deformation of the cells. At 150 °C, foam sheets maintaining the bead structure are formed when the molding time is relatively short at 2–3 min; however, at longer molding times, deformation of the sheet surface can be seen contrary to those molded at 145 °C. Surface and cross section SEM micrographs of the samples, prepared under the same conditions as in Figure 5, are shown in Figure 6. The surface of the foam sheet molded at 145 °C in Figure 6a is smooth and does not show irregular surface melting of the TPU, but that molded at 150 °C in Figure 6b shows irregular surface melting and consequently, destructive deformation of the surface. It seems that similar cell morphology was obtained between the core and the close-to-skin layer. Under both conditions, the interface between the beads becomes thicker with molding time, suggesting effective sintering of the bead surfaces. Although there is no deformation of the cell structure when molded at 145 °C, cell deformation from the original ETPU occurs at 150 °C with an increase in molding time due to melting, especially at the interface between beads where interfacial sintering occurs.

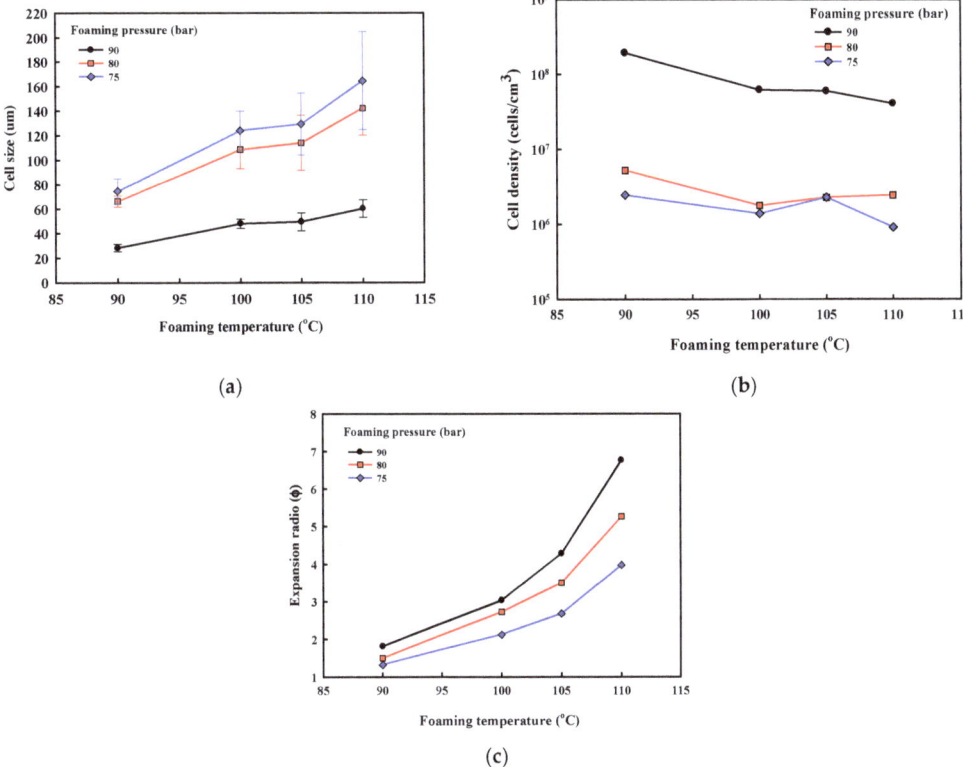

**Figure 3.** Physical properties of ETPU beads prepared at different foaming temperatures and pressures in the supercritical $CO_2$ foaming process: (**a**) foam size; (**b**) foam density; (**c**) expansion ratio.

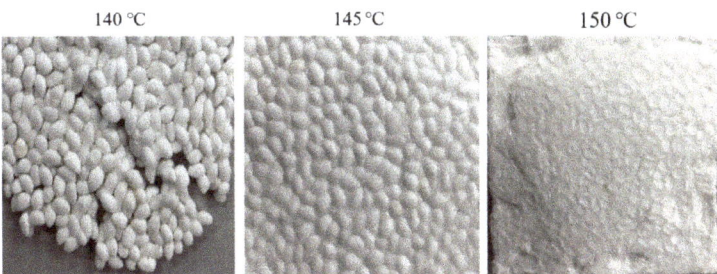

**Figure 4.** Effect of molding temperature in the compression molding of ETPU beads at 3.5 MPa for 15 min.

The effect of compression molding pressure on the sintering of beads is shown in Figure 7. The interface between beads becomes thicker with the increase in pressure which may increase the physical properties of the foam sheets; however, destructive deformation of the foam surface and cell deformation near the interface can be seen as in the case of increasing molding times (Figure 6). Thus, compression molding at 3.5 MPa appears to result in the best foam sheets. Based on these results, compression molding of ETPU foam sheets is possible, but when compared with steam chest injection molding, the temperature and pressure range at which cell deformation can be minimized is very limited and thus, precise control of the molding temperature and pressure is required.

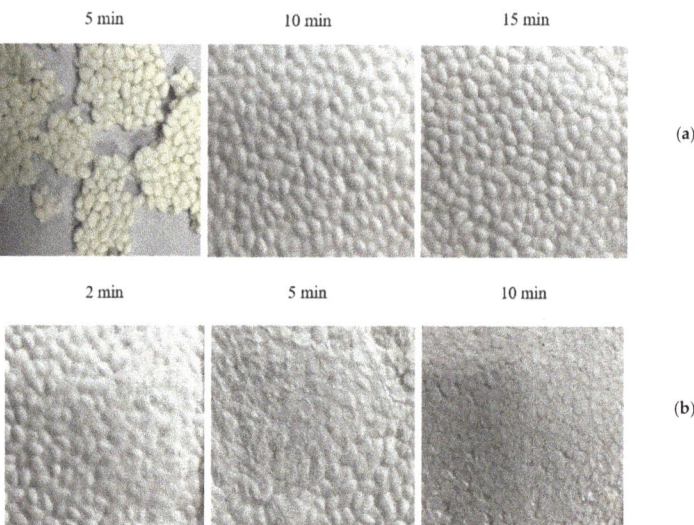

**Figure 5.** Effect of molding time in the compression molding of ETPU beads at 3.5 MPa, molding temperature: (**a**) 145 °C; (**b**) 150 °C.

The effect of cell structure on the properties of the foam sheets is studied by compression molding ETPU beads, prepared under different conditions and thus, having different cell structure, at 3.5 MPa and 145 °C for 15 min, which is the molding condition where sintering of the beads occurs and deformation of the cell can be minimized. The SEM micrographs of foam sheets compression molded at 145 °C for 15 min with ETPU having different cell diameter and cell density are shown in Figure 8. In all ETPU, sintering through surface fusion was sufficient; when the ETPU foamed at low pressure and temperature is used, the interface formed by fusion of the beads is thicker and the cell structure is not deformed in the compression molding process.

The effect of the cell diameter, cell density, and expansion ratio on the rebound properties of the compression molded foam sheets is shown in Figure 9. The ball rebound property is generally used to evaluate the resilience of foams. Unlike hardness, the ball rebound property reflects the instantaneous feel of the foam and when the foam has poor resilience or low energy absorption, it exhibits lower rebound. The rebound property is generally controlled by appropriate selection of the isocyanate and polyol used in the polymerization of the polyurethane. However, as can be seen in Figure 9, even with a single polyurethane different ball, rebound properties can be obtained by compression molding ETPU of different cell structure, obtained by foaming TPU under different conditions. The ball rebound property is dependent on the expansion ratio and is higher in the case of foams having higher expansion ratios (Figure 9), suggesting that medium-density foams have higher foam resilience and energy absorption compared with high-density foams. The foam sheet prepared with ETPU that foamed at 75 bar exhibits a relatively low ball rebound (Figure 9), which appears to be due to the insufficient cell expansion in TPU at the low foaming pressure (Figure 2). The theoretical expansion ratio that can be calculated from the cell volume ($V_g$), which, in turn, can be calculated from the number of cells ($N$) and the cell diameter ($D$) in Figure 3 and the theoretical expansion ratio ($\Phi_{Theoretical\ value}$) from the pellet volume ($V_p = 1$), is shown in Figure 9, along with the measured data. It shows a similar correlation with the experimental expansion ratio ($\Phi$) calculated using the measured densities before and after foaming, suggesting the theoretical expansion ratio calculated considering the two mutually complementary factors—cell diameter and cell density—correlates with the properties of the foam sheet.

$$Vg = N\frac{\pi}{6}D^3 \tag{3}$$

$$\Phi_{\text{Theoretical value}} = (V_p + V_g)/V_p = 1 + V_g = 1 + N\frac{\pi}{6}D^3 \tag{4}$$

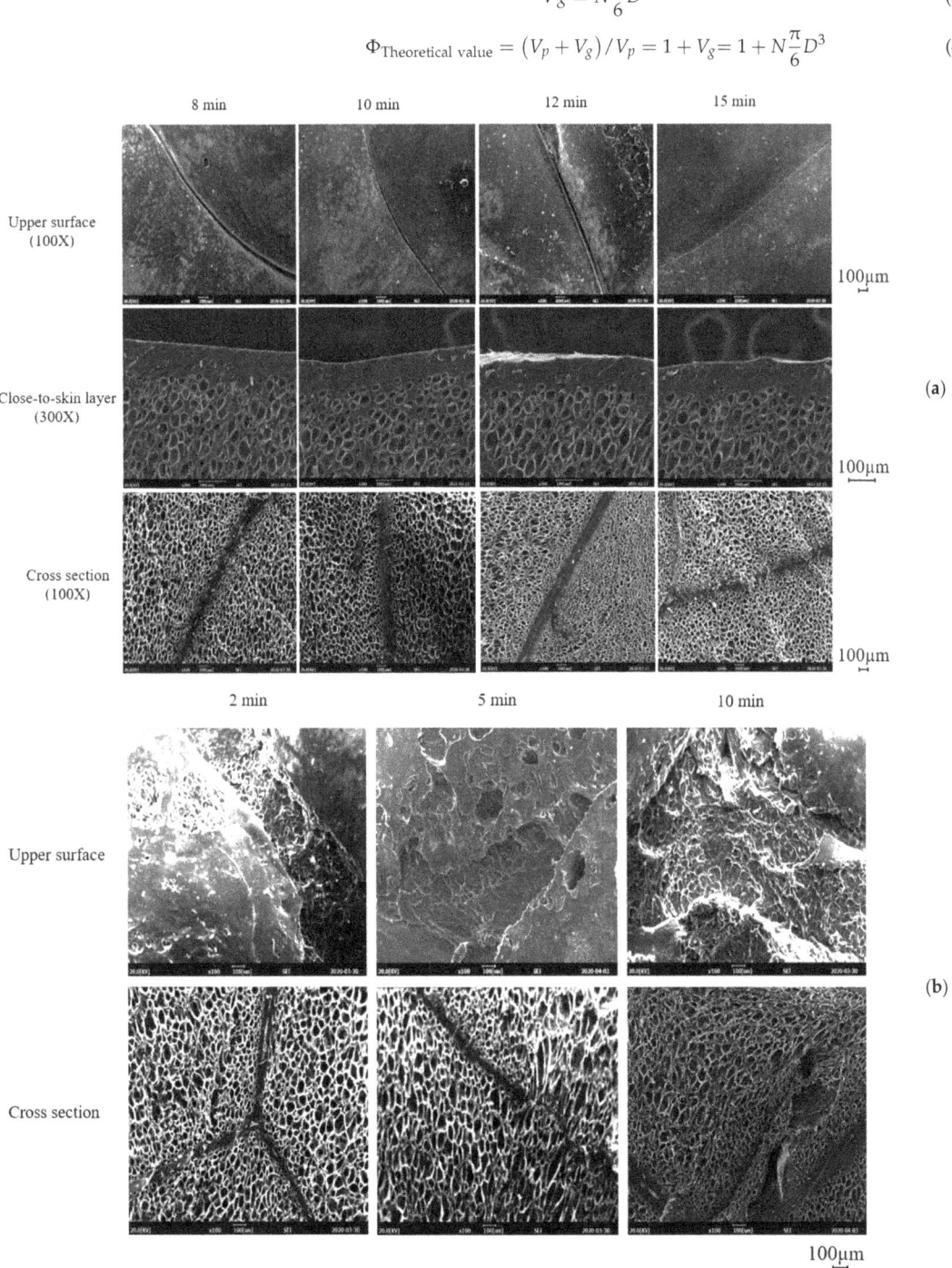

**Figure 6.** SEM micrographs of the surface and cross section of ETPU sheets molded at (**a**) 145 °C and (**b**) 150 °C.

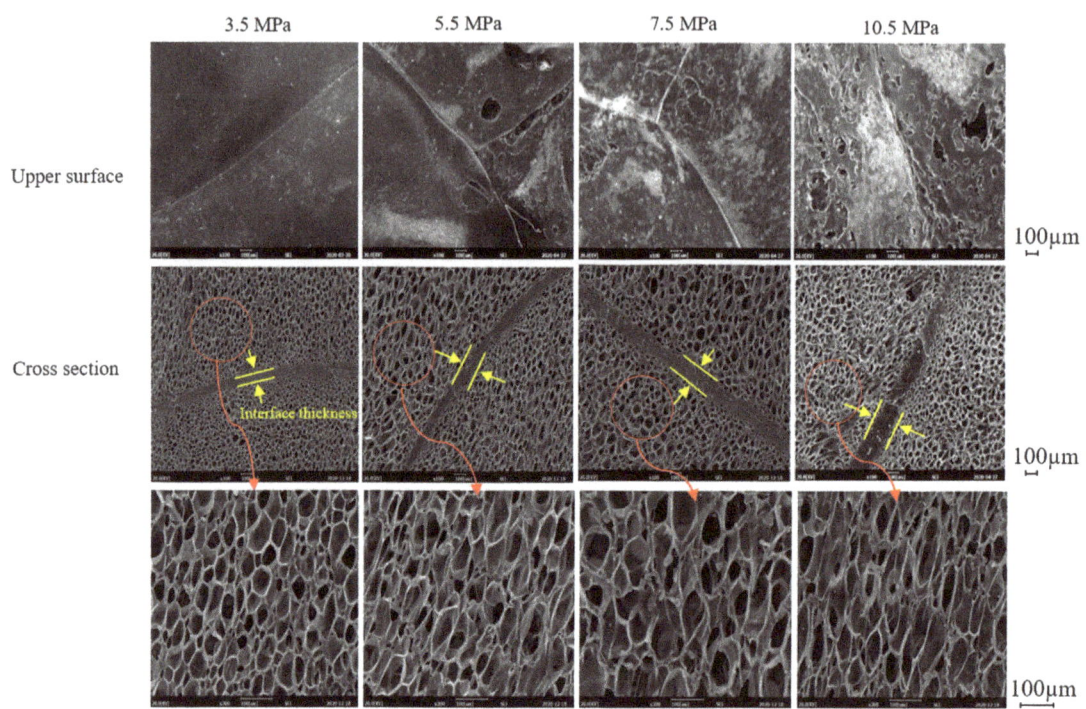

**Figure 7.** Effect of compression molding pressure on the sintering of ETPU beads at 145 °C.

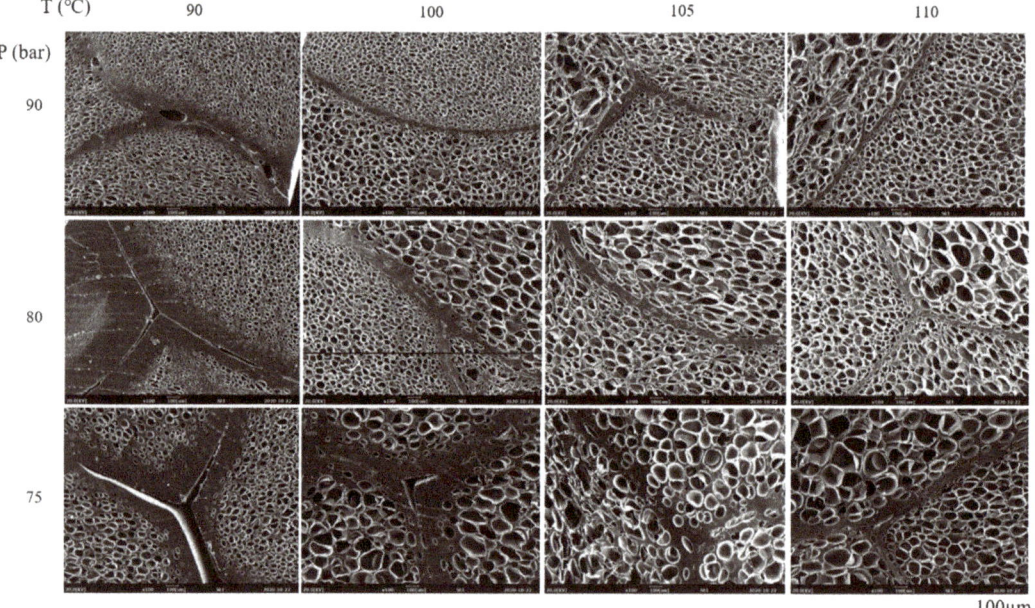

**Figure 8.** SEM micrographs of cross section of ETPU foam sheets made by compression molding at 3.5 MPa and 145 °C for 15 min.

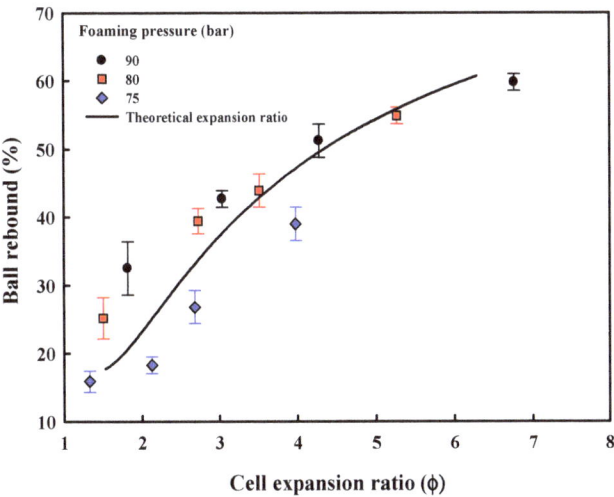

**Figure 9.** Effect of foam structure on the ball rebound of ETPU foam sheets.

Figure 10 shows the Shore A hardness of the prepared foam sheets. Contrary to the ball rebound property, foams with a low expansion ratio due to small cell diameter and a small number of cells were relatively hard; on the other hand, foams with higher expansion ratios exhibited low hardness and thus, were soft. The foam sheet prepared with ETPU foamed at low pressure and temperature of 75 bar and 90 °C, where cell expansion was not complete, shows a hardness value similar to the TPU sheet which had not been foamed. This appears to be due to the inadequate foaming condition resulting in a greater portion of the ETPU not being foamed. The hardness showed negligible change with the increase in the expansion ratio above 4, showing that in the case of the medium-density foam sheets (expansion ratio $\geq 4$), the expansion ratio determined by the number and size of cells does not affect the hardness of the foam sheets, while it does in the case of high-density foams (expansion ratio < 4).

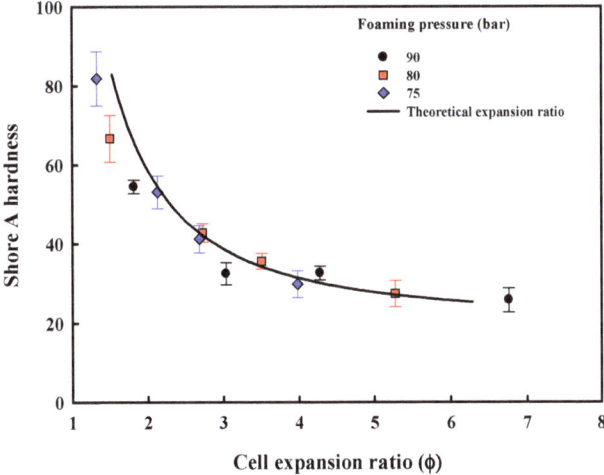

**Figure 10.** Effect of foam structure on Shore A hardness of ETPU foam sheets.

The tensile properties of the foam sheets are shown in Figure 11. The tensile strength and elongation at break increase with the increase in the expansion ratio but the modulus decreases. Foam sheets made from ETPU beads foamed at a relatively low pressure and temperature with thick intercell walls and thus, low expansion ratios are ruptured easily at the sintered interface between the beads by the applied tensile force, as the modulus of the parts that have not been foamed is higher. The tensile strength of the foam sheets made from expanded beads is dependent on the failure of the sintered interface between the beads and the failure of the cells inside the beads. When the compression molding conditions are not adequate and sintering is insufficient, failure at the bead interface is expected to result in very poor tensile strengths, but the mechanical properties of the foam sheets processed under appropriate conditions are expected to depend on the failure at the interface or cell depending on the cell structure. In Figure 12, showing the cross-section SEM micrographs of fracture surfaces resulting from tensile testing of the foam sheets processed under optimum compression molding conditions in this study, it can be observed that failure at both the interface and cells occurs with the relative degree depending on the ETPU used. Foam sheets made from low expansion ratio beads fail at the bead–bead interface, while those having higher expansion ratios from higher foaming pressures and temperatures fail at the cell, resulting in higher tensile strengths and elongation at break. That is, the closed cell structure in the beads with high expansion ratios absorbs the energy in tensile testing without failure at the interface until the cell finally fails instead of the interface. Based on these results, it appears that using medium-density foam beads rather than high-density foam beads is advantageous for adequate mechanical strengths of foam sheets made by compression molding.

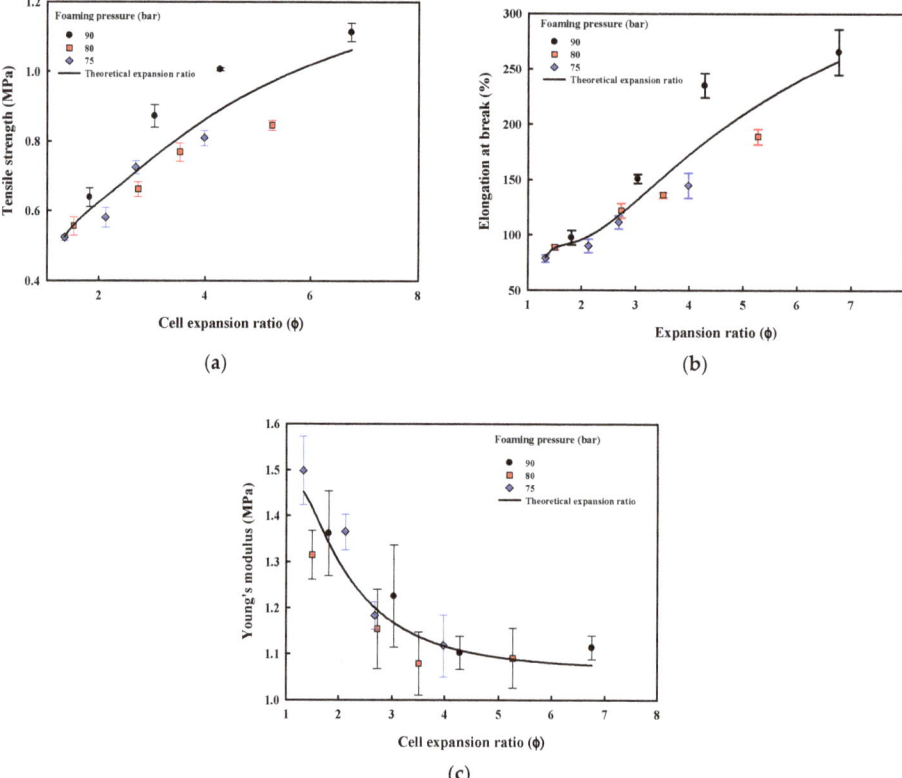

**Figure 11.** Mechanical properties of ETPU foam sheets: (**a**) tensile strength; (**b**) elongation at break; (**c**) Young's modulus.

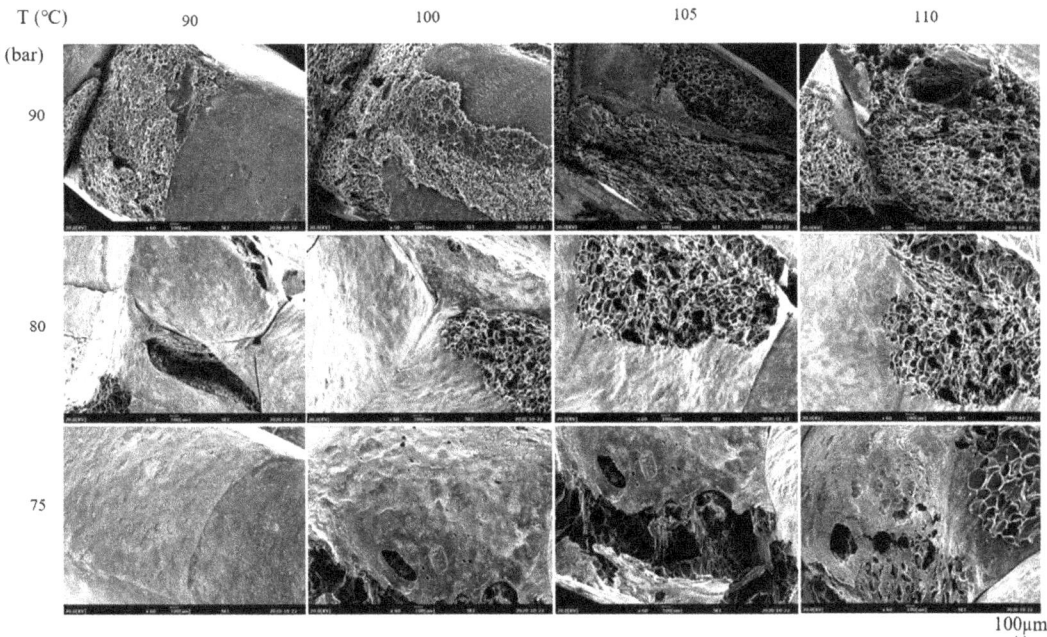

**Figure 12.** SEM micrographs of the fracture surface of ETPU foam sheets from tensile testing.

## 4. Conclusions

The effect of the foaming pressure and temperature on cell formation in expanded TPU and the possibilities of preparing foam sheets by compression molding the prepared EPTU were studied. The effect of the structure of the foamed beads on the properties of the compression molded foam sheet was studied to obtain the following conclusion.

The TPU used in this study exhibits closed cell structures when foamed at 75–90 bar and 90–110 °C and ETPU beads having diverse foam structure with fine and/or conventional cells can be made. At higher foaming pressures, more nuclei are formed and the cell size decreases, resulting in higher expansion ratios, and the increase in the foaming temperature at a fixed pressure affects the viscoelastic property to increase the cell size and decrease the number of cells, resulting in lower expansion ratios. The possibilities of compression molding ETPU to prepare foam sheets have been confirmed, but the processing window to obtain foam sheets where the cell structure in the EPTU is not deformed during sintering of ETPU beads is very narrow. The expansion ratio of the ETPU affects the foam sheet properties with higher expansion ratios, resulting in lower hardness and thus, higher resilience. The developed cell structure also contributes to higher mechanical properties such as tensile strength and elongation at break.

**Author Contributions:** Conceptualization, H.-J.K.; Investigation, T.Z., S.-J.L., Y.H.Y., K.-H.P.; Writing—original draft, T.Z., H.-J.K. All authors have read and agreed to the published version of the manuscript.

**Funding:** This research was funded by Gyeonggi-do through the Gyeonggi-do Regional Research Center (GRRC) Program (Project: Development of functional microcomposite materials for precision molding of flexible materials, GRRC Dankook 2016-B02).

**Institutional Review Board Statement:** Not applicable.

**Informed Consent Statement:** Not applicable.

**Data Availability Statement:** Data are in the authors' possession.

**Conflicts of Interest:** The authors declare no conflict of interest.

## References

1. Mills, N.J. Chapter 1–Introduction to polymer foam microstructure. In *Polymer Foams Handbook: Engineering and Biomechanics Application and Design Guide*, 1st ed.; Butterworth-Heinemannn Ltd.: Oxford, UK, 2007; pp. 1–18.
2. Han, X.M.; Zeng, C.C.; Lee, L.J.; Koelling, K.W.; Tomasko, D.L. Extrusion of Polystyrene Nanocomposite Foams With Supercritical $CO_2$. *Poly. Eng. Sci.* **2003**, *43*, 1261–1275. [CrossRef]
3. Hayashi, H.; Mori, T.; Okamoto, M.; Yamasaki, S.; Hayami, H. Polyethylene ionomer-based nano-composite foams prepared by a batch process and MuCell® injection molding. *Mater. Sci. Eng. C* **2010**, *30*, 62–70. [CrossRef]
4. Gomez-Monterde, J.; Hain, J.; Sanchez-Soto, M.; Maspoch, M.L. Microcellular injection moulding: A comparison between MuCell process and the novel micro-foaming technology IQ Foam. *J. Mater. Process. Technol.* **2019**, *268*, 162–170. [CrossRef]
5. Raps, D.; Hossieny, N.; Park, C.B.; Altstadt, V. Past and present developments in polymer bead foams and bead foaming technology. *Polymer* **2015**, *56*, 5–19. [CrossRef]
6. Nofar, M.; Ameli, A.; Park, C.B. A novel technology to manufacture biodegradable polylactide bead foam products. *Mater. Des.* **2015**, *83*, 413–421. [CrossRef]
7. Kang, D.H.; Oh, S.S.; Kim, H.I. Improvement of Physical Properties of Polypropylene Chemical Foam by Glass Fiber Reinforcement. *Polym. Korea* **2019**, *43*, 589–594. [CrossRef]
8. Li, Q.; Matuana, L.M. Foam extrusion of high density polyethylene/wood-flour composites using chemical foaming agents. *J. Appl. Polym. Sci.* **2003**, *28*, 3139–3150. [CrossRef]
9. Shin, J.H.; Lee, H.K.; Song, K.B.; Lee, K.H. Characterization of Poly(lactic acid) Foams Prepared with Supercritical Carbon Dioxode. *Polym. Korea* **2013**, *37*, 685–693. [CrossRef]
10. Sun, Y.; Usda, Y.; Suganaga, H.; Haruki, M.; Kihara, S.; Takishima, S. Pressure drop threshold in the foaming of low density polyethylene, polystyrene, and polypropylene using $CO_2$ and $N_2$ as foaming agents. *J. Supercrit. Fluid.* **2015**, *103*, 38–47. [CrossRef]
11. He, J.; Gao, Q.; Song, X.; Bu, X.; He, J. Effect of foaming agent on physical and mechanical properties of alkali-activated slag foamed concrete. *Constr. Build. Mater.* **2019**, *226*, 280–287. [CrossRef]
12. Zhang, R.; Huang, K.; Hu, S.F.; Liu, Q.T.; Zhao, X.; Liu, Y. Improved cell morphology and reduced shrinkage ratio of ETPU beads by reactive blending. *Polym. Test.* **2017**, *63*, 38–46. [CrossRef]
13. Standau, T.; Hadelt, B.; Schreier, P.; Altstadt, V. Development a Bead Foam from an Engineering Polymer with Addition of Chain Extender: Expanded Polybutylene Terephthalate. *Ind. Eng. Chem. Res.* **2018**, *57*, 17170–17176. [CrossRef]
14. Shen, J.; Cao, X.; Lee, L.J. Synthesis and foaming of water expandable polystyrene-clay nanocomposites. *Polymer* **2006**, *47*, 6303–6310. [CrossRef]
15. Wali, K.F.; Bhavnani, H.; Overfelt, R.A.; Sheldon, D.S.; Williams, K. Investigation of the Performance of an Expandable Polystyrene Injector for Use on the Lost-Foam Casting Process. *Metall. Mater. Trans. B* **2003**, *34*, 843–851. [CrossRef]
16. Ruiz, J.A.R.; Vincent, M.; Agassant, J.F.; Sadik, T.; Caroline, P. Polymer foaming with chemical blowing agents: Experiment and modeling. *Polym. Eng. Sci.* **2015**, *55*, 2018–2029. [CrossRef]
17. Hossieny, N.J.; Barzegari, M.R.; Nofar, M.; Mahmood, S.H.; Park, C.B. Crystallization of hard segment domains with the presence of butane for microcellular thermoplastic polyurethane foams. *Polymer* **2014**, *55*, 651–662. [CrossRef]
18. Guo, Y.; Hossieny, N.; Chu, R.K.M.; Park, C.B.; Zhou, N. Critical processing parameters for foamed bead manufacturing in a lab-scale autoclave system. *Chem. Eng. J.* **2013**, *214*, 180–188. [CrossRef]
19. Naduib, H.E.; Park, C.B.; Reichelt, N. Fundamental Foaming Mechanisms Governing the Volume Expansion of Extruded Polypropylene Foams. *J. Appl. Polym. Sci.* **2004**, *91*, 2661–2668.
20. Ji, W.; Wang, D.; Guo, J.; Fei, B.; Gu, X.; Li, H.; Sun, J.; Zhang, S. The preparation of starch derivatives reacted with urea-phosphoric acid and effect on fire performance of expandable polystyrene foams. Carbohydr. *Polymers* **2020**, *233*, 115841.
21. Nofar, M.; Guo, Y.; Park, C.B. Double Crystal Melting Peak Generation for Expanded Polypropylene Bead Foam Manufacturing. *Ind. Eng. Chem. Res.* **2013**, *52*, 2297–2303. [CrossRef]
22. Sulong, N.H.R.; Mustapa, S.A.S.; Rashid, M.K.A. Application of expanded polystyrene (EPS) in buildings and constructions: A review. *J. Appl. Polym. Sci.* **2019**, *136*, 47529–47539. [CrossRef]
23. Guo, P.; Xu, Y.; Lu, M.; Zhang, S. Expanded Linear Low- Density Polyethylene Beads: Fabrication, Melt Strength, and Foam Morphology. *Ind. Eng. Chem. Res.* **2016**, *55*, 8104–8113. [CrossRef]
24. Zhao, J.; Wang, G.; Wang, C.; Park, C.B. Ultra-lightweight, super thermal-insulation and strong PP/CNT microcellular foams. *Compos. Sci. Technol.* **2020**, *191*, 108084–108095. [CrossRef]
25. Srivastava, V.; Srivastava, R. A Review on Manufacturing, Properties and Application of Expanded Polypropylene. *MIT. Int. J. Mech. Eng.* **2014**, *4*, 22–28.
26. Ge, C.B.; Wang, S.P.; Zheng, W.G.; Zhai, W.T. Preparation of Microcellular Thermoplastic Polyurethane (TPU) Foam and Its Tensile Property. *Polym. Eng. Sci.* **2018**, *58*, E158–E166. [CrossRef]
27. Jiang, X.; Zhao, L.; Feng, L.; Chen, C. Microcellular Thermoplastic Polyurethanes and their flexible properties prepared by mold foaming process with supercritical $CO_2$. *J. Cell. Plast.* **2019**, *55*, 615–631. [CrossRef]
28. Zhai, W.T.; Kim, Y.W.; Jung, D.W.; Park, C.B. Steam-Chest Molding of Expanded Polypropylene Foams. 2. Mechanism of Interbead Bonding. *Ind. Eng. Chem. Res.* **2011**, *50*, 5523–5531. [CrossRef]

29. Hossieny, N.; Ameli, A.; Park, C.B. Characterization of Expanded Polypropylene Bead Foams with Modified Steam-Chest Molding. *Ind. Eng. Chem. Res.* **2013**, *52*, 8236–8247. [CrossRef]
30. Ge, C.B.; Een, Q.; Wang, S.P.; Zheng, W.G.; Zhai, W.T.; Park, C.B. Steam-chest molding of expanded thermoplastic polyurethane bead foams and their mechanical properties. *Chem. Eng. Sci.* **2017**, *174*, 337–346. [CrossRef]
31. Zhao, D.; Wang, G.J.; Wang, M.H. Investigation of the effect of foaming process parameters on expanded thermoplastic polyurethane bead foams properties using response surface methodology. *J. Appl. Polym. Sci.* **2018**, *135*, 46327–46337. [CrossRef]
32. Okolieocha, C.; Raps, D.; Subramanianm, K.; Altstadt, V. Microcellular to nanocellular polymer foams: Progress (2004–2015) and future directions-A review. *Eur. Polym. J.* **2015**, *73*, 500–519. [CrossRef]